SPSS Missing Value Analysis™ 7.5

MaryAnn Hill / SPSS Inc.

SPSS Inc.
444 N. Michigan Avenue
Chicago, Illinois 60611
Tel: (312) 329-2400
Fax: (312) 329-3668

SPSS Federal Systems (U.S.)
SPSS Argentina srl
SPSS Asia Pacific Pte. Ltd.
SPSS Australasia Pty. Ltd.
SPSS Belgium
SPSS Benelux BV
SPSS Central and Eastern Europe
SPSS East Mediterranea and Africa
SPSS France SARL
SPSS Germany
SPSS Hellas SA
SPSS Hispanoportuguesa S.L.
SPSS Ireland
SPSS Israel Ltd.
SPSS Italia srl
SPSS Japan Inc.
SPSS Korea
SPSS Latin America
SPSS Malaysia Sdn Bhd
SPSS Mexico SA de CV
SPSS Middle East and South Asia
SPSS Scandinavia AB
SPSS Schweiz AG
SPSS Singapore Pte. Ltd.
SPSS UK Ltd.

For more information about SPSS® software products, please visit our WWW site at http://www.spss.com or contact:

Marketing Department
SPSS Inc.
444 North Michigan Avenue
Chicago, IL 60611
Tel: (312) 329-2400
Fax: (312) 329-3668

SPSS is a registered trademark and the other product names are the trademarks of SPSS Inc. for its proprietary computer software. No material describing such software may be produced or distributed without the written permission of the owners of the trademark and license rights in the software and the copyrights in the published materials.

The SOFTWARE and documentation are provided with RESTRICTED RIGHTS. Use, duplication, or disclosure by the Government is subject to restrictions as set forth in subdivision (c)(1)(ii) of The Rights in Technical Data and Computer Software clause at 52.227-7013. Contractor/manufacturer is SPSS Inc., 444 N. Michigan Avenue, Chicago, IL, 60611.

General notice: Other product names mentioned herein are used for identification purposes only and may be trademarks of their respective companies.

TableLook is a trademark of SPSS Inc.
Windows is a registered trademark of Microsoft Corporation.

SPSS Missing Value Analysis™ 7.5
Copyright © 1997 by SPSS Inc.
All rights reserved.
Printed in the United States of America.

No part of this publication may be reproduced, stored in a retrieval system, or transmitted, in any form or by any means, electronic, mechanical, photocopying, recording, or otherwise, without the prior written permission of the publisher.

2 3 4 5 6 7 8 9 0   03 02 01 00 99 98 97

ISBN 1-56827-165-4

Library of Congress Catalog Card Number: 96-072435

Preface

SPSS is a powerful software package for data management and analysis. The Missing Value Analysis option extends this power by giving you tools for discovering patterns of missing data that occur frequently in survey and other types of data, and for dealing with data that contains missing values. Often in survey data, patterns become evident that will affect analysis. For example, you might find that people living in certain areas are reluctant to give their annual incomes, thus creating missing values in your data. If you leave these values out, are your statistical conclusions valid? Another concern is that the statistical algorithms being used make valid assumptions about the data distribution. With the Missing Value option, you can:

- Describe patterns of missing data.
- Estimate means, standard deviations, and correlations using a listwise, pairwise, regression, or EM (expectation-maximization) method.
- Fill in (impute) missing values with estimates obtained using a regression or an EM method.

The Missing Value procedure must be used with the SPSS Base system and is completely integrated into that system. You can use results from this procedure (for example, a correlation matrix) in other SPSS procedures.

About This Manual

This manual first presents the operation of the dialog box interface for Missing Value Analysis. The operational section is followed by extensive examples illustrating applications to various data configurations. Finally, the Syntax Reference section provides complete command syntax for the MVA command.

Compatibility

SPSS Missing Value Analysis, release 7.5, is designed to operate on computer systems running either Windows 95 or Windows NT 3.51.

Serial Numbers

Your serial number is your identification number with SPSS Inc. You will need this serial number when you call SPSS Inc. for information regarding support, payment, or an upgraded system. The serial number appears on the second diskette of systems that are distributed on diskette.

Registration Card

Don't put it off: *fill out and send us your registration card*. Until we receive your registration card, you have an unregistered system. Even if you have previously sent a card to us, please fill out and return the card enclosed in your Base system package. Registering your system entitles you to:
- Technical support services
- New product announcements and upgrade announcements

Customer Service

If you have any questions concerning your shipment or account, contact your local SPSS office, listed on p. vi. Please have your serial number ready for identification when calling.

Training Seminars

SPSS Inc. provides both public and onsite training seminars for SPSS. All seminars feature hands-on workshops. SPSS seminars will be offered in major U.S. and European cities on a regular basis. For more information on these seminars, call your local SPSS office, listed on p. vi.

Technical Support

The services of SPSS Technical Support are available to registered customers of SPSS. Customers may contact Technical Support for assistance in using SPSS products or for installation help for one of the supported hardware environments. To reach Technical Support, see the SPSS home page on the World Wide Web at http://www.spss.com, or call your local SPSS office, listed on p. vi. Be prepared to identify yourself, your organization, and the serial number of your system.

Additional Publications

Additional copies of SPSS product manuals may be purchased from Prentice Hall, the distributor of SPSS publications. To order, fill out and mail the Publications order form included with your system, or call toll-free. If you represent a bookstore or have an account with Prentice Hall, call 1-800-223-1360. If you are not an account customer, call 1-800-374-1200. In Canada, call 1-800-567-3800. Outside of North America, contact your local Prentice Hall office.

Except for academic course adoptions, manuals can also be purchased from SPSS Inc. Contact your local SPSS office, listed on p. vi.

Tell Us Your Thoughts

Your comments are important. Please send us a letter and let us know about your experiences with SPSS products. We especially like to hear about new and interesting applications using the SPSS system. Write to SPSS Inc. Marketing Department, Attn: Director of Product Planning, 444 N. Michigan Avenue, Chicago, IL 60611.

Contacting SPSS Inc.

If you would like to be on our mailing list, contact one of our offices below. We will send you a copy of our newsletter and let you know about SPSS Inc. activities in your area.

SPSS Inc.
Chicago, Illinois, U.S.A.
Tel: 1.312.329.2400
Fax: 1.312.329.3668
Customer Service:
1.800.521.1337
Sales:
1.800.543.2185
sales@spss.com
Training:
1-800-543-6607
Technical Support:
1.312.329.3410
support@spss.com

SPSS Federal Systems
Arlington, Virginia, U.S.A.
Tel: 1.703.527.6777
Fax: 1.703.527.6866

SPSS Argentina srl
Buenos Aires, Argentina
Tel: +541.816.4086
Fax: +541.814.5030

SPSS Asia Pacific Pte. Ltd.
Singapore, Singapore
Tel: +65.3922.738
Fax: +65.3922.739

SPSS Australasia Pty. Ltd.
Sydney, Australia
Tel: +61.2.9954.5660
Fax: +61.2.9954.5416

SPSS Belgium
Heverlee, Belgium
Tel: +32.162.389.82
Fax: +32.1620.0888

SPSS Benelux BV
Gorinchem, The Netherlands
Tel: +31.183.636711
Fax: +31.183.635839

SPSS Central and Eastern Europe
Woking, Surrey, U.K.
Tel: +44.(0)1483.719200
Fax: +44.(0)1483.719290

SPSS East Mediterranea and Africa
Herzelia, Israel
Tel: +972.9.526700
Fax: +972.9.526715

SPSS France SARL
Boulogne, France
Tel: +33.1.4699.9670
Fax: +33.1.4684.0180

SPSS Germany
Munich, Germany
Tel: +49.89.4890740
Fax: +49.89.4483115

SPSS Hellas SA
Athens, Greece
Tel: +30.1.7251925
Fax: +30.1.7249124

SPSS Hispanoportuguesa S.L.
Madrid, Spain
Tel: +34.1.443.3700
Fax: +34.1.448.6692

SPSS Ireland
Dublin, Ireland
Tel: +353.1.66.13788
Fax: +353.1.661.5200

SPSS Israel Ltd.
Herzlia, Israel
Tel: +972.9.526700
Fax: +972.9.526715

SPSS Italia srl
Bologna, Italy
Tel: +39.51.252573
Fax: +39.51.253285

SPSS Japan Inc.
Tokyo, Japan
Tel: +81.3.5474.0341
Fax: +81.3.5474.2678

SPSS Korea
Seoul, Korea
Tel: +82.2.552.9415
Fax: +82.2.539.0136

SPSS Latin America
Chicago, Illinois, U.S.A.
Tel: 1.312.494.3226
Fax: 1.312.494.3227

SPSS Malaysia Sdn Bhd
Selangor, Malaysia
Tel: +603.704.5877
Fax: +603.704.5790

SPSS Mexico SA de CV
Mexico DF, Mexico
Tel: +52.5.575.3091
Fax: +52.5.575.3094

SPSS Middle East and South Asia
Dubai, UAE
Tel: +971.4.525536
Fax: +971.4.524669

SPSS Scandinavia AB
Stockholm, Sweden
Tel: +46.8.102610
Fax: +46.8.102550

SPSS Schweiz AG
Zurich, Switzerland
Tel: +41.1.201.0930
Fax: +41.1.201.0921

SPSS Singapore Pte. Ltd.
Singapore, Singapore
Tel: +65.2991238
Fax: +65.2990849

SPSS UK Ltd.
Woking, Surrey, U.K.
Tel: +44.1483.719200
Fax: +44.1483.719290

Contents

1 Missing Value Analysis 1

To Obtain a Missing Value Analysis 2
    Missing Value Analysis Patterns 3
    Missing Value Analysis Descriptives 5
    Missing Value Analysis Regression 6
    Missing Value Analysis EM 7
    Missing Value Analysis Variables for EM and Regression 8
    MVA Command Additional Features 9

2 Missing Data: Descriptive Displays, Estimates of Statistics, and Imputation of Values 11

Example 1:
A First Look at Patterns of Incompleteness 14

Example 2:
Pursuing Patterns Further 21

Example 3:
Patterns in a Large Survey 29

Example 4:
Estimating Means, Standard Deviations, Covariances, and Correlations 41

Example 5:
Estimating Replacement Values: Imputation 53

Syntax Reference

    MVA 63

Bibliography 77

Subject Index 79

Syntax Index 81

Missing Value Analysis

The Missing Value procedure performs three primary functions:
- Describes the pattern of missing data: where the missing values are located, how extensive they are, whether pairs of variables tend to have values missing in different cases, whether data values are extreme, and whether values are missing randomly.
- Estimates means, standard deviation, covariances, and correlations using a listwise, pairwise, regression, or EM (expectation-maximization) method. The pairwise method also displays counts of pairwise complete cases.
- Fills in (imputes) missing values with estimated values using regression or EM methods.

Missing value analysis helps address several concerns caused by incomplete data. Cases with missing values that are systematically different from cases without missing values can obscure the results. Also, missing data may reduce the precision of calculated statistics because there is less information than originally planned. Another concern is that the assumptions behind many statistical procedures are based on complete cases, and missing values can complicate the theory required.

Example. In evaluating a treatment for leukemia, several variables are measured. However, not all measurements are available for every patient. The patterns of missing data are displayed, tabulated, and found to be random. An EM analysis is used to estimate the means, correlations, and covariances. Missing values are replaced by imputed values and saved into a new data file to be used for further analysis.

Statistics. Univariate statistics, including number of nonmissing values, mean, standard deviation, number of missing values, and number of extreme values. Estimated means, covariance matrix, and correlation matrix, using listwise, pairwise, EM, or regression methods. Little's MCAR test with EM results. Summary of means by various methods. For groups defined by missing versus nonmissing values: t tests. For all variables: missing value patterns displayed cases-by-variables.

Data. Data can be categorical or quantitative. For each variable, missing values that are not coded as system-missing must be defined as user-missing. For example, if a questionnaire item has the response *Don't know* coded as 5 and you want to treat it as missing, the item should have 5 coded as a user-missing value.

Assumptions. Listwise and pairwise estimation depends on the assumption that the pattern of missing values does not depend on the data values. (This condition is known as

missing completely at random, or MCAR.) Violation of this assumption can lead to biased estimates. Regression and EM estimation depend on the assumption that the pattern of missing data is related to the observed data only. (This condition is called **missing at random**, or MAR.) This assumption allows estimates to be adjusted using available information.

Related procedures. Many procedures in SPSS allow you to use listwise or pairwise estimation. Linear Regression and Factor allow replacement of missing values by the mean values. In the SPSS Trends option, several methods are available to replace missing values in time series. To code user-missing values, choose *Define Variable* from the Data menu.

To Obtain a Missing Value Analysis

▶ From the menus choose:

Statistics
  Missing Value Analysis...

Figure 1.1 Missing Value Analysis dialog box

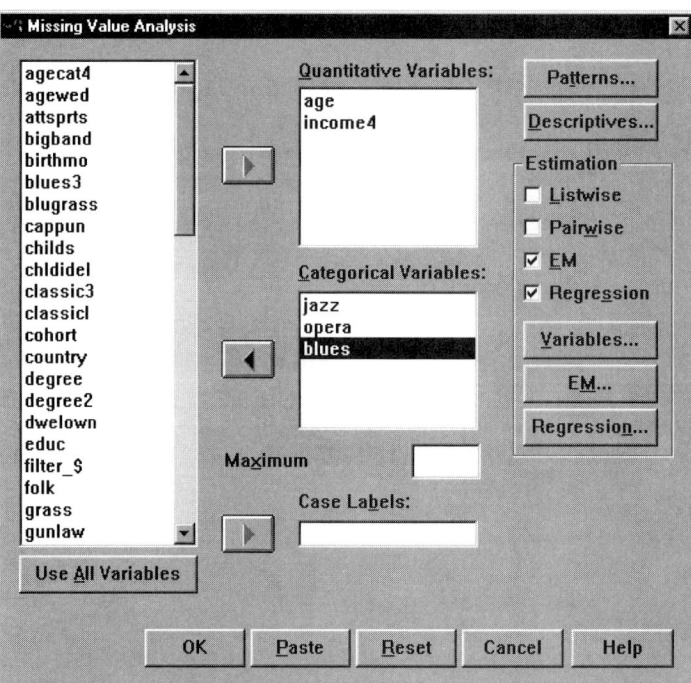

▶ Select at least one quantitative variable.

Optionally, you can:

- Select categorical variables (numeric or string) and enter a limit on the number of categories (*Maximum*).
- Click *Patterns* or *Descriptives* for descriptions of missing values.
- Select a method for estimation of statistics and estimation of the missing values themselves.
- If you select EM or Regression, click *Variables* to specify a subset to be used for the estimation.

Missing Value Analysis Patterns

Figure 1.2   Missing Value Analysis Patterns dialog box

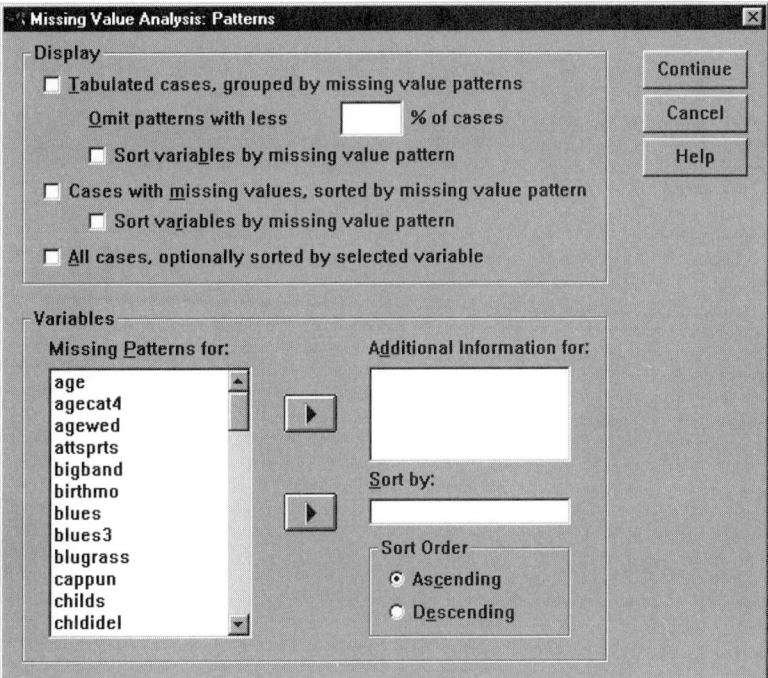

Display. Three types of pattern tables are available, containing cases or numbers of cases versus variables. Instead of values or counts, the cells of the table contain symbols that indicate the type of value. For *Tabulated cases*, *X*'s are used to indicate missing values. For *All cases* and *Cases with missing values*, the symbols in the display are:

+	Extremely high value
–	Extremely low value
S	System-missing value
A	First type of user-missing value
B	Second type of user-missing value
C	Third type of user-missing value

- **Tabulated cases.** The frequency of each missing value pattern is tabulated. Counts and variables are both sorted by similarity of patterns.
    Omit patterns with less than n % of cases. Eliminates patterns that occur infrequently.
- **Cases with missing values.** Case-by-variable patterns of missing and extreme values are shown only for cases that have missing values. Cases and variables are both sorted by similarity of patterns.
- **All cases.** For each case, the pattern of missing values and extreme values is displayed. Unless a sort variable is specified, cases are listed in the order in which they appear in the data file.

Variables. You can specify variables for labeling and sorting the pattern displays.
- **Missing Patterns for.** Lists all quantitative and categorical variables from the Missing Value Analysis dialog box.
- **Additional information for.** Lists values for each case. For tabulated patterns, this option lists the mean of quantitative variables or, for categorical variables, the number of cases having the pattern in each category.
- **Sort by.** Cases are listed according to the ascending or descending order of the values of the specified variable. Available only for *All cases*.

Missing Value Analysis Descriptives

Figure 1.3 Missing Value Analysis Descriptives dialog box

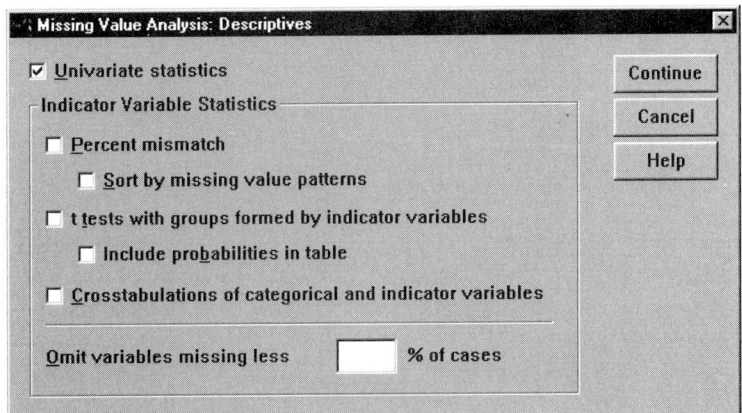

Univariate statistics. For each variable, displays the number of nonmissing values, the mean, the standard deviation, and the number and percentage of missing values. Also displays counts and percentages of missing values and counts of extremely high and low values. (Means, standard deviation, and extreme value counts are not reported for categorical variables.)

Indicator Variable Statistics. For each variable, SPSS creates a missing indicator variable that indicates whether the value of the variable is present or missing. The indicator variables are not displayed but are used in creating the mismatch, t test, and frequency tables. To reduce table size, you can omit statistics that are computed for only a small number of cases.

- **Percent mismatch.** For each pair of variables, displays the percentage of cases in which one variable has a missing value and the other variable has a nonmissing value. Each diagonal element in the table contains the percentage of missing values for a single variable.

- **t tests with groups formed by indicator variables.** The means of two groups are compared for each quantitative variable, using Student's *t* statistic. The groups are determined by whether the indicator variable is coded present or missing. The *t* statistic, degrees of freedom, counts of missing and nonmissing values, and means of the two groups are displayed. You can also display any two-tailed probabilities associated with the *t* statistics, although interpretation of these probabilities can be problematic.
- **Crosstabulations of categorical and indicator variables.** A table is displayed for each categorical variable. For each category, the table shows the frequency and percentage of nonmissing values for the other variables. The percentages of each type of missing value are also displayed.

Missing Value Analysis Regression

Figure 1.4 Missing Value Analysis Regression dialog box

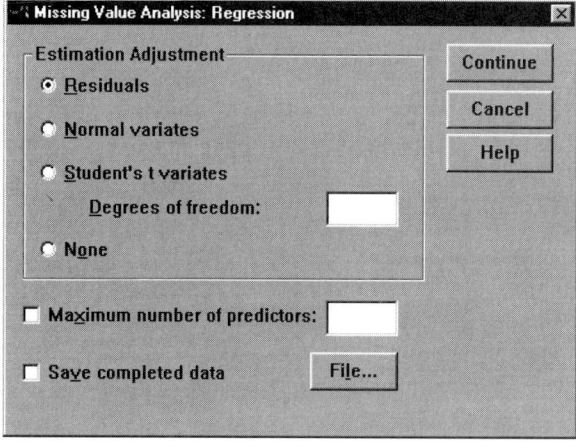

Regression estimates missing values using multiple linear regression. The means, the covariance matrix, and the correlation matrix of the predicted variables are displayed.

Estimation Adjustment. The regression method can add a random component to regression estimates. You can select residuals, normal variates, Student's *t* variates, or no adjustment.

Maximum number of predictors. Sets a maximum limit on the number of predictor (independent) variables used in the estimation process.

Save completed data. Writes an SPSS data file, with missing values replaced by values estimated by the regression method.

Missing Value Analysis EM

Figure 1.5 Missing Value Analysis EM dialog box

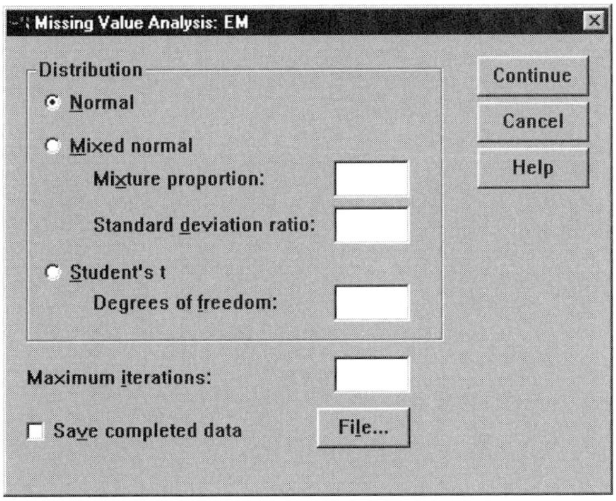

EM estimates the means, the covariance matrix, and the correlation of quantitative variables with missing values, using an iterative process.

Distribution. Various assumptions can be made for distribution of data: normal, mixed normal, and Student's t. For a mixed normal assumption, you can specify the proportion and the standard deviation ratio. For Student's t distribution, you must specify the degrees of freedom.

Maximum iterations. Sets the maximum number of iterations. The procedure stops when this number of iterations is reached, even if the estimates have not converged.

Save completed data. Writes an SPSS data file, with missing values replaced by values estimated by the EM method.

Missing Value Analysis Variables for EM and Regression

Figure 1.6   Missing Value Analysis Variables for EM and Regression dialog box

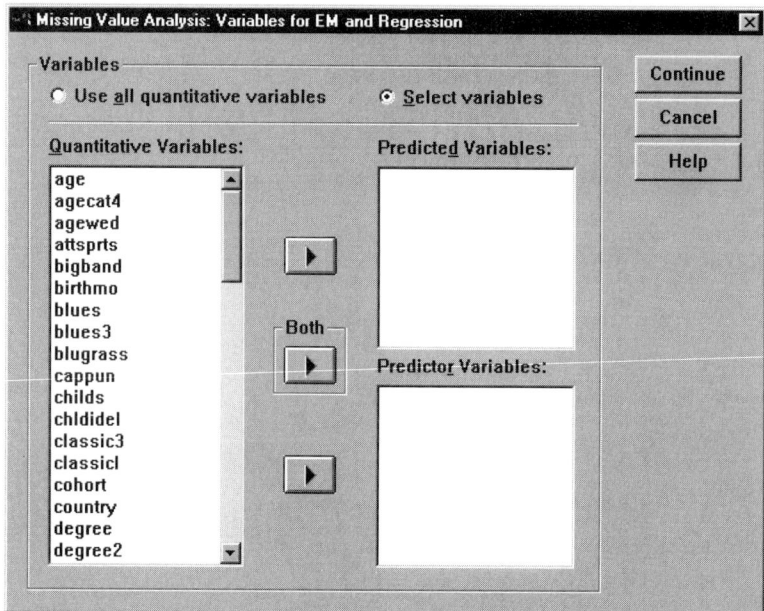

Variables. Quantitative variables were selected in the Missing Value Analysis dialog box. Variables in the Categorical list are not available here. Quantitative variables can be copied to either Predicted Variables, Predictor Variables, or both.

MVA Command Additional Features

The SPSS command language also allows you to:
- Specify separate descriptive variables for missing value patterns, data patterns, and tabulated patterns using the DESCRIBE keyword on the MPATTERN, DPATTERN, or TPATTERN subcommands.
- Specify more than one sort variable for the data patterns table, using the DPATTERN subcommand.
- Specify tolerance and convergence, using the EM subcommand.
- Specify tolerance and F-to-enter, using the REGRESSION subcommand.
- Specify different variable lists for EM and Regression, using the EM and REGRESSION subcommands.
- Specify different percentages for suppressing cases displayed for TTEST, TABULATE, and MISMATCH.

See the Syntax Reference section of this manual for complete syntax information.

2 Missing Data: Descriptive Displays, Estimates of Statistics, and Imputation of Values

Even in the best designed and monitored study, observations can be missing—a subject inadvertently skips a question, a blood sample is ruined, or the recording equipment malfunctions. Because many classical statistical analyses require complete cases (no missing values), when data are incomplete it may be hard "to get off the ground." That is, if the analyst wants to explore a new data set by, say, using a factor analysis to identify redundant variables or sets of related variables, a cluster analysis to check for distinct subpopulations, or a stepwise discriminant analysis to see which variables differ among subgroups, there may be too few complete cases for an analysis. For example, there are *no* complete cases in the survey data with 61 variables and 1500 cases, described below.

The features in the Missing Value procedure address three tasks:

- **Description of patterns.** How many missing values are there? Where are they located (specific cases and/or variables)? Are values missing randomly? For each variable, the word *pattern* indicates the dichotomized version of the variable—that is, a binary distribution where each value is *missing* or *present*. Also, when the same variables are missing for several cases, cases are said to have the same *pattern*.
- **Estimation of means, standard deviations, covariances, and correlations.** Statistics are computed using one or more methods: all values, listwise, pairwise, EM (expectation maximization), or regression. Several options are available for both the EM and regression methods.
- **Imputation of values.** EM and regression methods are provided for estimating replacement values for the missing data.

Methods for estimation and imputation are defined in the examples that follow. To us, none of the approaches should be viewed as a magic black box when the data are non-randomly incomplete. While the EM and regression methods allow a specific way in which the values of one variable may be related to another, a good data analyst will want to ferret out possible problems in how the data are sampled, recorded, or otherwise fail to conform to the study protocol—for example, which regions of a multivariate space are sparse because data are missing? It is hard to separate the selection of an appropriate method for estimation or imputation from the basic data screening process.

Often it is necessary to run the Missing Value procedure several times. You should:
- First, see the extent and pattern of missing values, and determine if values are missing randomly. At this point, you may want to delete cases and variables with large numbers of missing data and, most importantly, screen variables with skewed distributions for symmetrizing transformations before proceeding to the estimation or imputation phases.
- Next, study various estimates of descriptive statistics, possibly making a side step to check relations graphically when differences in estimates are found.
- Finally, impute values (estimate replacement values) and use graphics to assess the suitability of the filled-in values.

The use of a data matrix with imputed values may not be acceptable for a final report of results, but by using the approaches and methods described here, you may be able to find a subset of variables with enough complete cases for a meaningful analysis. You may omit variables simply because a large proportion of their values are missing; or, by making exploratory runs using the imputed data matrix, you may learn that some variables are redundant or have little relation to the outcome variables of interest. For example:
- In a stepwise regression, you may find that some variables have no relation to your outcome variable. Try rerunning the analysis with a smaller subset of candidate variables that has many more complete cases.
- In a factor analysis, you may identify one or more redundant variables. You might also learn this by examining an estimate of the correlation matrix in the MVA procedure.

Data. Two data sets are used in the examples below. We approach these data as consultants contacted after the data were collected and recorded, not as collaborators involved in designing a study.

For each of 109 countries in the *world95m* data file, 22 variables were culled from several 1995 almanacs—including life expectancy, birth rate, the ratio of birth rate to death rate, infant mortality, gross domestic product per capita, female and male literacy rates, average calories consumed per day, the percentage of the population living in cities, and so on. Because the values of some variables differed across sources, we are unsure of their accuracy. Four percent of the data values are missing, yet only 58 cases are complete (53% of the sample).

The General Social Survey (GSS) data contain a subset of 61 responses collected by the National Opinion Research Center at the University of Chicago in 1993 for 1500 people, age 18 years or older. Variables include the respondent's income in 1991 (*rincom91*), the general household income (*income91*), years of school attendance (*educ*), sex, view of life (*dull*, *routine*, or *exciting*), opinions about music (*jazz*, *classical*, *rap*), and so on. Eighteen percent of the values are missing and there are *no* complete cases.

Example 1: A first look at patterns of incompleteness. After a univariate summary report shows that the variable *calories* in the *world95m* data has the most values missing, representations, or pictures, of the data matrix displaying one character per variable are introduced (for example, an *S* represents a system-missing variable, and a blank represents a value that is present). In one variation, cases and variables are ordered by the pattern of incompleteness; in another, counts of cases with the most common patterns are reported. A tally of low and high values for each variable provides a clue that the distributions of three variables are skewed and need to be re-expressed.

Example 2: Pursuing patterns further. Using the missing value pattern of calories and also male and female literacy rates as grouping variables, two-sample t tests are requested for all quantitative variables. The same patterns are also crosstabulated against the categorical variables *region2* and *religion*. As a check on whether pairs of variables tend to have values missing for the same cases or whether they are mismatched (if one is missing, the other is present), pairwise counts of values present and percentages of pairwise mismatched cases are reported.

Example 3: Patterns in a large survey. In the 1993 General Social Survey, three subsets of items are administered to different subsamples of two-thirds of the cases, so when items are selected from each subset no complete cases remain. Our codebook provides no identification of the subsets or subsamples. By using a display of the common patterns of missing data and information about pairwise mismatches, the subsets are identified. Many of the items in this survey have two or more codes for missing values. Boxplots and the crosstabulation of patterns against categorical variables highlight relations among distributions in the categories and multiple missing value codes.

Example 4: Estimating means, standard deviations, covariances, and correlations. For each quantitative variable, pairwise estimates of means and standard deviations are computed for each subsample formed by pairing the variable with each other variable. For easy comparison, estimates obtained by the listwise, all values, EM, and regression methods are displayed in a summary panel. Correlations are estimated using each of the default methods, and differences between elements in each pair of matrices are computed and displayed using SPSS's MATRIX procedure.

Example 5: Estimating replacement values: Imputation. Replacements for missing values are estimated by both the EM and regression methods. The filled-in data matrices are saved and the identity of the observed and imputed values displayed in scatterplots. Estimates from the two methods for the same variable are compared graphically.

Pivot table editing. In the examples below, we edited many of the SPSS output panels, requesting fewer digits following the decimal point, omitting results for variables that are similar to others, and so forth. Each task begins by clicking on the table to select it and selecting *SPSS Pivot Table Object* from the Edit menu (or double-click the table). Then:

- To omit cases (rows) or columns of a table, press the Ctrl and Alt keys and simultaneously click on the row(s) or columns(s) to highlight them, and then choose *Clear* or *Delete* from the Edit menu.
- To control the width of cells and digits following the decimal, first choose *Select/Table body* on the Edit menu and then *Cell Properties* or *Set Data Cell Width* from the Format menu.
- To transpose rows and columns, select *Transpose rows and columns* from the Pivot menu.
- To move elements in cells into "layers," select *Pivoting Trays* from the Pivot menu and drag the icons.
- To rotate column labels from horizontal to vertical or vice-versa, choose *Rotate Inner Column Labels* from the Format menu.

Example 1:
A First Look at Patterns of Incompleteness

Where are the missing values located? How extensive are they? If a value is missing for one variable, does it tend to be missing for one or more other variables? Conversely, if a value is present for one variable, do values tend to be missing for other specific variables? Is the pattern of missing values related to values of another variable?

You may need to uncover patterns of incomplete data in order to:

- select enough complete cases for a meaningful analysis. If you omit a few variables, or even just one, does the sample size of complete cases increase dramatically?
- select a method of estimation or imputation. If, for example, you plan to use complete cases for a final analysis, you need to verify that values are missing *completely* at random.
- understand how results may be biased or distorted because of a failure to meet necessary assumptions about randomness of the missing values.

Three items from the GSS survey described above form a good illustration of variables that can not be used together in a multivariate procedure. The survey has three subsets of questions administered to two-thirds of the sample. It is possible to include one item from each set, with the result that no cases are complete. For example, look at the crosstabulation of the patterns for questions about gun laws, marijuana, and euthanasia:

Table 2.1 Crosstabulation of patterns that shows no complete cases

GUNLAW	GRASS	LETDIE1 missing	LETDIE1 present
missing	missing	7	27
	present	34	448
present	missing	55	481
	present	448	0

The sample sizes for the items are, respectively, 984, 930, and 956, and yet no subject answered all three questions. The percents of pairwise mismatched cases are 65%, 68%, and 66%! From the tabulated pattern display in Example 3, it is clear that the large blocks of incomplete data occur for different subsets of the cases.

A study of burn victims provides an example of possibly distorted results. If the ultimate goal is a model to predict survival for burn victims using age, blood gasses, total area burned, etc., as predictors, many would find it disconcerting that a sizable portion of the healthiest part of the sample is missing—only 31% of the survivors had blood gas determinations, but more than 90% of those who died had them. In addition, because of pediatric hospital unit policy, blood gasses were measured for only 20% of the children under the age of 7.

In this example, we explore the *world95m* data for patterns of how values are missing. It begins to be clear in several displays that values of female and male literacy are not missing randomly. In Example 2, we continue the search and description of patterns after log transforming some variables with skewed distributions.

To produce this output, from the menus choose:

Statistics
  Missing Value Analysis...

▸ Quantitative Variables: populatn, density, urban, lifeexpf, lifeexpm, literacy, pop_incr, babymort, gdp_cap, calories, birth_rt, death_rt, log_gdp, lg_aidsr, b_to_d, fertilty, log_pop, lit_male, lit_fema

▸ Categorical Variables: region2, religion, climate

▸ Case Labels: country

Patterns...

  Display
  ☑ Tabulated cases, grouped by missing value patterns
  ☑ Cases with missing values, sorted by missing value patterns
  ☑ All cases, optionally sorted by selected variable

  Variables
  ☑ Additional information: region2, literacy, religion, babymort, gdp_cap
  ☑ Sort by: region2

Univariate statistics. This panel provides your first look, variable by variable, at the extent of incomplete data. The number of values present is reported in the second column; the number missing, in the fifth column; and the percentage missing, in the sixth column. For *calories*, 75 countries (cases) report a value, and 34 do not. That is, *calories* is missing for 31.2% of the cases. The female and male literacy rates (*lit_fema* and *lit_male*) are each missing for 22% of the cases. Ten variables have no missing values, and nine others have from 0.9% to 2.8% missing values.

Because means and standard deviations are computed using all available data for each variable, the sample sizes vary from variable to variable. Statistics are not computed for *region2, religion,* and *climate* because they are specified as categorical variables.

Univariate Statistics

	N	Mean	Std. Deviation	Missing Count	Missing Percent	No. of Extremes[a] Low	No. of Extremes[a] High
POPULATN	109	47723.88	146726.36	0	.0	0	11
DENSITY	109	203.415	675.705	0	.0	0	13
URBAN	108	56.53	24.20	1	.9	0	0
LIFEEXPF	109	70.16	10.57	0	.0	9	0
LIFEEXPM	109	64.92	9.27	0	.0	6	0
LITERACY	107	78.34	22.88	2	1.8	0	0
POP_INCR	109	1.682	1.198	0	.0	0	0
BABYMORT	109	42.313	38.079	0	.0	0	1
GDP_CAP	109	5859.98	6479.84	0	.0	0	13
CALORIES	75	2753.83	567.83	34	31.2	0	0
BIRTH_RT	109	25.923	12.361	0	.0	0	0
DEATH_RT	108	9.56	4.25	1	.9	0	9
LOG_GDP	109	3.4218	.6201	0	.0	0	0
LG_AIDSR	106	1.3800	.7094	3	2.8	0	0
B_TO_D	108	3.2035	2.1250	1	.9	0	2
FERTILTY	107	3.563	1.902	2	1.8	0	0
LOG_POP	109	4.1140	.6542	0	.0	2	2
LIT_MALE	85	78.73	20.45	24	22.0	0	0
LIT_FEMA	85	67.26	28.61	24	22.0	0	0
REGION2	107			2	1.8		
RELIGION	108			1	.9		
CLIMATE	107			2	1.8		

a. Number of cases outside the range (Q1 - 1.5*IQR, Q3 + 1.5*IQR).

For each variable, Tukey's robust boxplot criterion for "outside values" is used to define extreme values. (For larger files the criterion is the sample mean plus or minus two standard deviations.) The extremes are tabulated in the last two columns. Use these counts to identify possible outliers and/or skewed distributions. Symmetry is important if one's goal is to estimate means, standard deviations, covariances, or correlations. More than 10% of the population, density, and GDP per capita values are *high* (the counts are 11,

13, and 13) and none are *low*. Boxplots of these distributions show that they are right-skewed; thus, for estimation and imputation, we log-transform them.

The first three boxplots in Figure 2.1 use the data as recorded. In the last three boxplots, each variable is log transformed. In order to display the six distributions within a single frame, the variables are transformed to z scores before plotting. The shape of each distribution is emphasized because the maximum value is set as 4.0, eliminating the outliers China and India for *population* and Hong Kong and Singapore for *density* in the first three boxplots.

Figure 2.1 Boxplots of population, density, and GDP per capita before and after log transformation

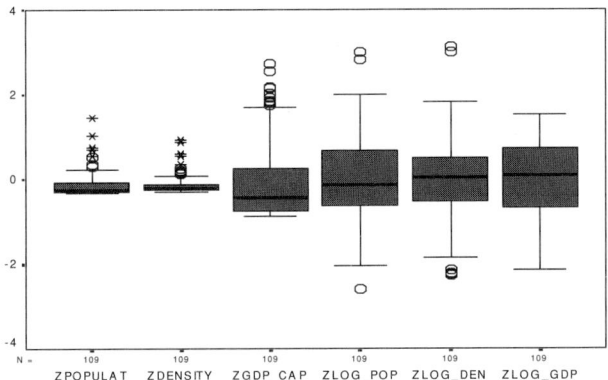

Casewise patterns of incomplete data. The display Data Patterns (all cases) is a picture of the data file that highlights the location of missing observations and extreme values. Each column in the display represents the values of a variable; each row represents the data for one country or subject. (To save space, the Eastern European, African, and Latin American countries are omitted.) Countries are ordered by geographical region, because *region2* is specified as the Sort variable. The values of *region2* for Canada and the USA are missing, so they are listed first.

A single print character is allowed for each variable: a blank when the value is present and not extreme, a plus sign (+) for an extremely large value, a minus sign (−) for an extremely small value, an *S* for system missing (for example, a blank in the data), and up to three user-defined missing codes, denoted by *A*, *B*, and *C*. This display is used to see if particular cases and/or variables have too little complete data to use and also to see if variables (or groups of variables) have values missing nonrandomly.

Data Patterns (all cases)

Case	# Missing	% Missing	POPULATN	DENSITY	URBAN	LIFEEXPF	LIFEEXPM	LITERACY	POP_INCR	BABYMORT	GDP_CAP	CALORIES	BIRTH_RT	DEATH_RT	LOG_GDP	LG_AIDSR	B_TO_D	FERTILTY	LOG_POP	LIT_MALE	LIT_FEMA	REGION2	RELIGION	CLIMATE	LITERACY	BABYMORT	GDP_CAP	REGION2	RELIGION
Canada	3	13.6							+											S	S	S			97	6.8	19904	.	Catholic
USA	1	4.5	+						+													S			97	8.1	23474	.	Protstnt
Austria	2	9.1							+											S	S				99	6.7	18396	Europe	Catholic
Belgium	3	13.6		+					+	S										S	S				99	7.2	17912	Europe	Catholic
Denmark	2	9.1							+											S	S				99	6.6	18277	Europe	Protstnt
Finland	2	9.1																		S	S				100	5.3	15877	Europe	Protstnt
France	2	9.1							+											S	S				99	6.7	18944	Europe	Catholic
Germany	2	9.1							+											S	S				99	6.5	17539	Europe	Protstnt
Greece	0	.0																							93	8.2	8060	Europe	Orthodox
Iceland	3	13.6							+	S							-			S	S				100	4.0	17241	Europe	Protstnt
Ireland	2	9.1																		S	S				98	7.4	12170	Europe	Catholic
Italy	0	.0							+																97	7.6	17500	Europe	Catholic
Netherlands	2	9.1		+					+											S	S				99	6.3	17245	Europe	Catholic
Norway	2	9.1							+											S	S				99	6.3	17755	Europe	Protstnt
Portugal	1	4.5								S															85	9.2	9000	Europe	Catholic
Spain	0	.0																							95	6.9	13047	Europe	Catholic
Sweden	2	9.1																		S	S				99	5.7	16900	Europe	Protstnt
Switzerland	2	9.1							+											S	S				99	6.2	22384	Europe	Catholic
UK	2	9.1																		S	S				99	7.2	15974	Europe	Protstnt
Afghanistan	1	4.5			-			+		S		+													29	168.0	205	Pacific/Asia	Muslim
Australia	0	.0																							100	7.3	16848	Pacific/Asia	Protstnt
Bangladesh	0	.0	+	+																					35	106.0	202	Pacific/Asia	Muslim
Cambodia	0	.0																								35	112.0	260	Pacific/Asia	Buddhist
China	0	.0	+														+								78	52.0	377	Pacific/Asia	Taoist
Hong Kong	1	4.5		+						S															77	5.8	14641	Pacific/Asia	Buddhist
India	0	.0	+	+													+								52	79.0	275	Pacific/Asia	Hindu
Indonesia	0	.0	+																						77	68.0	681	Pacific/Asia	Muslim
Japan	2	9.1	+	+					+											S	S				99	4.4	19860	Pacific/Asia	Buddhist
Malaysia	0	.0																							78	25.6	2995	Pacific/Asia	Muslim
N. Korea	1	4.5								S															99	27.7	1000	Pacific/Asia	Buddhist
New Zealand	2	9.1																		S	S				99	8.9	14381	Pacific/Asia	Protstnt
Pakistan	1	4.5	+							S															35	101.0	406	Pacific/Asia	Muslim
Philippines	0	.0																							90	51.0	867	Pacific/Asia	Catholic
S. Korea	1	4.5			+					S															96	21.7	6627	Pacific/Asia	Protstnt
Singapore	0	.0			+																				88	5.7	14990	Pacific/Asia	Taoist
Taiwan	8	36.4			+					S		S		S	S	S		S	S					S	91	5.1	7055	Pacific/Asia	Buddhist
Thailand	0	.0																							93	37.0	1800	Pacific/Asia	Buddhist
Vietnam	0	.0																						S		88	46.0	230	Pacific/Asia	Buddhist
Armenia	2	9.1								S														S		98	27.0	5000	Middle East	Orthodox
Azerbaijan	2	9.1								S					S											98	35.0	3000	Middle East	Muslim
Bahrain	1	4.5			+					S															77	25.0	7875	Middle East	Muslim
Egypt	0	.0																							48	76.4	748	Middle East	Muslim
Iran	0	.0																							54	60.0	1500	Middle East	Muslim
Iraq	0	.0																							60	67.0	1955	Middle East	Muslim
Israel	1	4.5								S															92	8.6	13066	Middle East	Jewish
Jordan	0	.0																							80	34.0	1157	Middle East	Muslim
Kuwait	0	.0																+							73	12.5	6818	Middle East	Muslim
Lebanon	1	4.5			+					S															80	39.5	1429	Middle East	Muslim
Libya	0	.0																							64	63.0	5910	Middle East	Muslim
Oman	4	18.2					S			S										S	S				.	36.7	7467	Middle East	Muslim
Saudi Arabia	0	.0																							62	52.0	6651	Middle East	Muslim
Syria	1	4.5								S															64	43.0	2436	Middle East	Muslim
Turkey	0	.0																							81	49.0	3721	Middle East	Muslim
U.Arab Em.	1	4.5								S							+								68	22.0	14193	Middle East	Muslim
Uzbekistan	1	4.5								S															97	53.0	1350	Middle East	Muslim

The *S*'s show that when female literacy is missing, male literacy is missing too. *Lit_male* and *lit_fema* are missing frequently for European countries, but *calories* is missing more often for Middle Eastern countries. In the complete sample, 36.4% of Taiwan's data are missing, 22.7% of Bosnia's data are missing (not shown), and so forth. A "+" indicates that China's and India's populations, for example, are extremely large. Some analysts recommend treating a value that is an extreme outlier (and not a recording error) as a missing value.

The user can opt to display values of variables with the patterns. Here, values of the categorical variables *region2* and *religion* are displayed along with the values of the quantitative variables literacy, infant mortality (*babymort*), and GDP. For the European countries, the literacy and GDP per capita values appear higher than those for most of the other countries, while infant mortality (*babymort*) is lower. Variability appears greatest within the Pacific/Asia region.

Sorted casewise patterns. In Missing Patterns (cases with missing values), cases and variables are sorted by the patterns of the missing data. The last three columns are *lit_male*, *lit_fema*, and *calories*, and the two last cases are Bosnia and Taiwan, because they have the most values missing. Complete cases are not included. To shorten the output, we omit countries with one missing value and no extreme values (*calories* is missing for most of the omitted cases).

Missing Patterns (cases with missing values)

Case	# Missing	% Missing	POPULATN	DENSITY	LIFEEXPF	LIFEEXPM	POP_INCR	BABYMORT	GDP_CAP	BIRTH_RT	LOG_GDP	LOG_POP	DEATH_RT	B_TO_D	FERTILTY	LG_AIDSR	CLIMATE	URBAN	LITERACY	RELIGION	REGION2	LIT_FEMA	LIT_MALE	CALORIES	LITERACY	BABYMORT	GDP_CAP	REGION2	RELIGION
Afghanistan	1	4.5		−			+				+													S	29	168.0	205	Pacific/Asia	Muslim
Bahrain	1	4.5																						S	77	25.0	7875	Middle East	Muslim
Barbados	1	4.5		+			−																	S	99	20.3	6950	Latn America	Protstnt
Hong Kong	1	4.5		+																				S	77	5.8	14641	Pacific/Asia	Buddhist
Lebanon	1	4.5		+																				S	80	39.5	1429	Middle East	Muslim
Pakistan	1	4.5	+																					S	35	101.0	406	Pacific/Asia	Muslim
Russia	1	4.5	+																					S	99	27.0	6680	East Europe	Orthodox
S. Korea	1	4.5		+																				S	96	21.7	6627	Pacific/Asia	Protstnt
U.Arab Em.	1	4.5														+								S	68	22.0	14193	Middle East	Muslim
Armenia	2	9.1															S							S	98	27.0	5000	Middle East	Orthodox
Azerbaijan	2	9.1													S									S	98	35.0	3000	Middle East	Muslim
USA	1	4.5	+																S						97	8.1	23474	.	Protstnt
Canada	3	13.6						+													S	S	S		97	6.8	19904	.	Catholic
Denmark	2	9.1						+														S	S		99	6.6	18277	Europe	Protstnt
Netherlands	2	9.1		+				+														S	S		99	6.3	17245	Europe	Catholic
New Zealand	2	9.1																				S	S		99	8.9	14381	Pacific/Asia	Protstnt
Norway	2	9.1						+														S	S		99	6.3	17755	Europe	Protstnt
Austria	2	9.1						+														S	S		99	6.7	18396	Europe	Catholic
Finland	2	9.1																				S	S		100	5.3	15877	Europe	Protstnt
France	2	9.1						+														S	S		99	6.7	18944	Europe	Catholic
Romania	2	9.1																				S	S		96	20.3	2702	East Europe	Orthodox
Japan	2	9.1	+	+				+														S	S		99	4.4	19860	Pacific/Asia	Buddhist
Sweden	2	9.1																				S	S		99	5.7	16900	Europe	Protstnt
Switzerland	2	9.1						+														S	S		99	6.2	22384	Europe	Catholic
Germany	2	9.1						+														S	S		99	6.5	17539	Europe	Protstnt
UK	2	9.1																				S	S		99	7.2	15974	Europe	Protstnt
Ireland	2	9.1																				S	S		98	7.4	12170	Europe	Catholic
Bulgaria	3	13.6																				S	S	S	93	12.0	3831	East Europe	Orthodox
Belgium	3	13.6		+				+														S	S	S	99	7.2	17912	Europe	Catholic
Croatia	3	13.6																				S	S	S	97	8.7	5487	East Europe	Catholic
Iceland	3	13.6						+		−												S	S	S	100	4.0	17241	Europe	Protstnt
Oman	4	18.2																S				S	S	S	.	36.7	7467	Middle East	Muslim
Czech Rep.	4	18.2																S	S			S	S	S	.	9.3	7311	East Europe	Catholic
South Africa	4	18.2																			A	S	S	S	76	47.1	3128	Africa	.
Bosnia	5	22.7									S	S										S	S	S	86	12.7	3098	East Europe	Muslim
Taiwan	8	36.4		+							S	S	S	S	S							S	S	S	91	5.1	7055	Pacific/Asia	Buddhist

It is easy to see that when *calories* is missing, the literacy rates tend to be present. For larger data files, the most common patterns may be less apparent; so, in the next display, common patterns are tabulated.

Common patterns of missing data. In the Tabulated Patterns display, variables are listed in the same order as in the Missing Pattern display, and the common patterns are tabulated. We hid columns for the first 10 variables because they are empty. The first row in the display represents the pattern for 58 cases and has blanks for all 22 variables—that is, 58 cases have no values missing. For 24 cases, *calories* is the only missing value; for 14 others, the male and female literacy rates are missing; and for 4 cases, all three variables are missing. The sum of 58, 24, 14, and 4 does not equal the total sample size (109) because patterns unique to a single case are not displayed. By default, the pattern is omitted if less than 1% of the cases have it, but you can change the percentage.

The column labeled *Complete if...* reports the number of complete cases if the variable(s) marked by X in that pattern are omitted. Thus, if *calories* is eliminated, the number of complete cases increases from 58 to 82; if only the male and female literacy rates are omitted, the number is 72; and if all three variables are removed, it jumps to 100.

Tabulated Patterns

# of cases	BABYMORT	B_TO_D	FERTILITY	LG_AIDSR	CLIMATE	URBAN	LITERACY	RELIGION	REGION2	LIT_FEMA	LIT_MALE	CALORIES	Complete if...	LITERACY	BABYMORT	GDP_CAP	Europe	East Europe	Pacific/Asia	Africa	Middle East	Latn America	Hindu	Jewish	Tribal	Muslim	Taoist	Animist	Catholic	Protstnt	Buddhist	Orthodox
58													58	69.1	58.6	2757	3	1	11	16	8	19	1	0	1	15	2	4	28	3	3	1
24												X	82	81.6	37.1	5272	1	8	5	2	6	2	0	1	0	9	0	0	4	4	2	4
14										X	X		72	98.8	7.5	16315	11	1	2	0	0	0	0	0	0	0	0	0	5	7	1	1
4										X	X	X	100	97.3	8.0	11118	2	2	0	0	0	0	0	0	0	0	0	0	2	1	0	1

For each pattern, the user can request frequency counts of categorical variables and/or means of quantitative variables. From the tabulation of *region2*, literacy rates are missing for at least 13 (11 + 2) European countries, while *calories* is missing for at least 10 (8 + 2) Eastern European countries. Infant mortality is much higher for the 58 complete cases (an average of 58.6 deaths during infancy for every 1000 live births) than for the countries where male and female literacy are missing (means of 7.5 and 8.0). The average *gdp_cap* for the 14 countries where *lit_fema* and *lit_male* are missing is $16,315, while it is only $2,757 for the 58 countries with no missing data. It is hard to believe that values are missing randomly!

Example 1:
Pursuing Patterns Further

In this example, we investigate the nonrandom pattern of incomplete data further by using the patterns of missing data for calories and the male and female literacy rates as grouping variables in two-sample *t* tests and also in crosstabulations with the categorical variables *region2* and *religion*. We also examine which pairs of variables have values that are jointly missing or mismatched (if one is present, the other is missing).

After viewing the output in Example 1, we decided that population, density, and GDP per capita should be re-expressed in log units. Here we transform *density* and omit the untransformed versions of the three variables.

To create this example, start by computing the log of each density value. From the menus choose:

Transform
  Compute...

log_den = LG10(density)

Recall the Missing Value Analysis dialog box from Example 1, and add *log_den* to the list of Quantitative variables. Remove *populatn*, *density*, and *gdp_cap*, and deselect the requests for the pattern displays shown in Example 1, as follows:

Statistics
  Missing Value Analysis...

▶ Quantitative Variables: urban, lifeexpf, lifeexpm, literacy, pop_incr, babymort, calories, birth_rt, death_rt, log_gdp, lg_aidsr, b_to_d, fertilty, log_pop, lit_male, lit_fema, log_den

▶ Categorical Variables: region2, religion, climate

▶ Case Labels: country
  Estimation
  ☑ Pairwise

Patterns...
  Display
  ☐ Tabulated cases, grouped by missing value patterns (deselect)
  ☐ Cases with missing values, sorted by missing value patterns (deselect)
  ☐ All cases, optionally sorted by selected variable (deselect)

Descriptives...
  Indicator Variable Statistics
  ☑ Percent mismatch
  ☑ *t* tests with groups formed by indicator variables
  ☑ Crosstabulations of categorical and indicator variables

Pairwise frequency counts. The Tabulated Patterns display in Example 1 on p. 21 provides one picture of the pattern of incomplete data, and a table of frequency counts for each pair of variables gives another view. This panel of pairwise frequencies is printed when you select *Pairwise* under Estimation in the Missing Value Analysis dialog box. Other results produced by this option are described in Example 4.

Pairwise Frequencies

	LIFEEXPF	LIFEEXPM	POP_INCR	BABYMORT	BIRTH_RT	LOG_GDP	LOG_POP	LOG_DEN	DEATH_RT	B_TO_D	FERTILTY	LG_AIDSR	CLIMATE	URBAN	LITERACY	RELIGION	REGION2	LIT_FEMA	LIT_MALE	CALORIES
LIFEEXPF	109																			
LIFEEXPM	109	109																		
POP_INCR	109	109	109																	
BABYMORT	109	109	109	109																
BIRTH_RT	109	109	109	109	109															
LOG_GDP	109	109	109	109	109	109														
LOG_POP	109	109	109	109	109	109	109													
LOG_DEN	109	109	109	109	109	109	109	109												
DEATH_RT	108	108	108	108	108	108	108	108	108											
B_TO_D	108	108	108	108	108	108	108	108	108	108										
FERTILTY	107	107	107	107	107	107	107	107	107	107	107									
LG_AIDSR	106	106	106	106	106	106	106	106	106	106	106	106								
CLIMATE	107	107	107	107	107	107	107	107	107	107	106	105	107							
URBAN	108	108	108	108	108	108	108	108	107	107	106	105	106	108						
LITERACY	107	107	107	107	107	107	107	107	106	106	105	104	105	107	107					
RELIGION	108	108	108	108	108	108	108	108	107	107	106	105	106	107	106	108				
REGION2	107	107	107	107	107	107	107	107	106	106	105	105	106	106	105	106	107			
LIT_FEMA	85	85	85	85	85	85	85	85	85	85	85	85	84	84	85	85	85	84	85	
LIT_MALE	85	85	85	85	85	85	85	85	85	85	85	85	84	84	85	85	85	84	85	85
CALORIES	75	75	75	75	75	75	75	75	75	75	75	75	75	74	74	75	73	59	59	75

The sample size for each variable is reported on the diagonal of the table; sample sizes for complete pairs of cases, off the diagonal. *Calories* alone has 75 values, but when paired with male or female literacy, the count of cases with both values drops to 59. If you need a set of variables for a multivariate analysis, it would be wise to omit calories or the male and female literacy rates. Otherwise, if these variables are essential to your analysis, be concerned that results may be biased due to the fact they are not missing randomly.

Pairwise mismatched patterns. For each pair of calorie and female literacy values, 50 cases (from the previous display, 109 total cases minus 59 cases that have both *lit_fema* and *calories* present) are not complete; both values or just one or the other may be missing (that is, the pattern is mismatched for the latter). If you need to omit variables in order to increase the number of complete cases, it helps to know which pairs of variables seldom occur together. From the Percent Mismatch of Indicator Variables table, the entry for *calories* paired with *lit_fema* is 38.5%. Almost 40% of the cases have either *calories* or a female literacy rate, but not both. The percentage missing for each variable individually is reported on the diagonal. If the percentage missing is less than 5%, the variable is not included in this display. You can change this cutoff for percentage missing in the Descriptives dialog box. Notice that here, by default, the variables are ordered by percentage missing. You can deselect this option on the same dialog box.

Percent Mismatch of Indicator Variables

	LIT_MALE	LIT_FEMA	CALORIES
LIT_MALE	22.02		
LIT_FEMA	.00	22.02	
CALORIES	38.53	38.53	31.19

Identifying nonrandom patterns with t tests. A two-sample *t* test is one way to check if data are **missing completely at random** (called MCAR by Little and Rubin). If the values of a variable are MCAR, then other quantitative variables should have roughly the same distribution for cases separated into two groups based on pattern (missing or present). In Figure 2.2, the pattern of female literacy groups infant mortality. Results of the *t* tests below confirm that average infant mortality is significantly higher when female literacy is present than when it is missing.

Figure 2.2   Boxplots of infant mortality grouped by pattern of female literacy

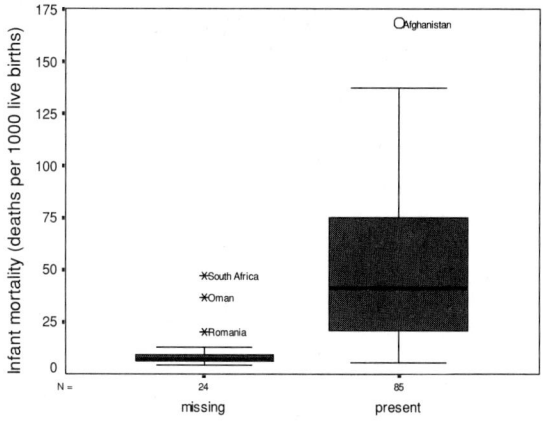

For each quantitative variable, the Separate Variances t Tests display has a column of two-sample t tests that compare its means for pattern variables (rows) with at least 5% missing data. (When only a few values are missing, t statistics are not informative.) The 5% limit can be changed in the Descriptives dialog box.

We show the t test results in two ways: first, the t statistics alone (the cell statistics are pivoted into layers) and then the default output layout (we hid *lifeexpm* and *lit_male*). For larger data files, it helps first to scan the display of t statistics looking for large values (positive or negative) and then refer to the complete table for more information.

Separate Variance t Tests

	URBAN	LIFEEXPF	LIFEEXPM	LITERACY	POP_INCR	BABYMORT	CALORIES	BIRTH_RT	DEATH_RT	LOG_GDP	LG_AIDSR	B_TO_D	FERTILTY	LOG_POP	LIT_MALE	LIT_FEMA	LOG_DEN
CALORIES	-1.2	-2.3	-2.0	-2.0	1.7	2.0	.	2.1	1.2	-2.0	4.5	.4	1.8	2.5	-2.3	-2.5	-2.2
LIT_MALE	-3.1	-7.8	-7.6	-8.1	6.7	8.7	-8.6	7.2	.3	-9.2	-1.8	4.5	5.7	.9		.	-.2
LIT_FEMA	-3.1	-7.8	-7.6	-8.1	6.7	8.7	-8.6	7.2	.3	-9.2	-1.8	4.5	5.7	.9	.		-.2

The t values greater than 2.0 or less than –2.0 are highlighted if you choose *Options* on the Edit menu, then choose *Scripts*, and select *MVA_Table_TOUT_MATRIXOFTSTATISTICS _Create*.

Both the size and location of the highlighted t's confirm our earlier observation that the literacy rates and calories have different nonrandom patterns of missing data. The absolute values of the t statistics are considerably larger for the literacy rates than for calories, so the departure from randomness is greater; and only 2 of the 11 variables with highlighted t's for the literacy rates are highlighted for calories (*lifeexpf* and *birth_rt*).

Separate Variance t Tests

		URBAN	LIFEEXPF	LITERACY	POP_INCR	BABYMORT	CALORIES	BIRTH_RT	DEATH_RT	LOG_GDP	LG_AIDSR	B_TO_D	FERTILTY	LOG_POP	LIT_FEMA	LOG_DEN
CALORIES	t	-1.2	-2.3	-2.0	1.7	2.0	.	2.1	1.2	-2.0	4.5	.4	1.8	2.5	-2.5	-2.2
	df	72.7	90.3	66.5	58.1	70.3	.	68.7	72	85.9	67.3	58.6	65.0	59.8	44.7	62.9
	# Present	74	75	74	75	75	75	75	75	75	75	75	75	75	59	75
	# Missing	34	34	33	34	34	0	34	33	34	31	33	32	34	26	34
	Mean(Present)	54.7	68.8	75	1.82	47.0	2754	27.5	9.9	3.35	1.55	3.26	3.8	4.22	62	1.70
	Mean(Missing)	60.6	73.1	85	1.38	31.9	.	22.3	8.9	3.58	.97	3.07	3.1	3.88	79	1.97
LIT_MALE	t	-3.1	-7.8	-8.1	6.7	8.7	-8.6	7.2	.3	-9.2	-1.8	4.5	5.7	.9	.	-.2
	df	40.9	105	105	49.9	106.9	53.7	64.7	90	73.4	63.7	47.8	54.2	41.2	.	35.6
	# Present	85	85	85	85	85	59	85	85	85	84	85	85	85	85	85
	# Missing	23	24	22	24	24	16	24	23	24	22	23	22	24	0	24
	Mean(Present)	53.2	67.8	74	1.99	51.3	2589	29.0	9.6	3.24	1.34	3.58	3.9	4.14	67	1.78
	Mean(Missing)	68.8	78.5	96	.610	10.6	3362	15.2	9.4	4.05	1.55	1.81	2.1	4.02	.	1.80
LIT_FEMA	t	-3.1	-7.8	-8.1	6.7	8.7	-8.6	7.2	.3	-9.2	-1.8	4.5	5.7	.9	.	-.2
	df	40.9	105	105	49.9	106.9	53.7	64.7	90	73.4	63.7	47.8	54.2	41.2	.	35.6
	# Present	85	85	85	85	85	59	85	85	85	84	85	85	85	85	85
	# Missing	23	24	22	24	24	16	24	23	24	22	23	22	24	0	24
	Mean(Present)	53.2	67.8	74	1.99	51.3	2589	29.0	9.6	3.24	1.34	3.58	3.9	4.14	67	1.78
	Mean(Missing)	68.8	78.5	96	.610	10.6	3362	15.2	9.4	4.05	1.55	1.81	2.1	4.02	.	1.80

Each cell in the second panel of separate variances *t* tests includes both the sample sizes and means of the *present* and *missing* groups, and the degrees of freedom for the separate variances two-sample *t* test. When the latter is markedly smaller than the sample size, the variances of the two groups differ. The sample size for the first five variables is 108 or 109, but the *df* for *pop_incr* are only 58. You can request the probability associated with each *t* statistic in the Descriptives dialog box. However, don't use these probabilities for significance testing because they are appropriate only when a single test is made and not in this situation of multiple tests.

Using the pattern of female literacy to separate the values of infant mortality into two groups, the *t* is 8.7. When female literacy is present, average *babymort* is 51.3 babies; when missing, it is 10.6 babies. The *t*'s are large for several other variables in the same row. Figure 2.3 shows a profile of these mean differences (before plotting, the variables were transformed to *z* scores). When female literacy is missing, the means of the first five variables (*b_to_d* through *birth_rt*) are lower than when it is present; the opposite is true for the last five variables (*urban* through *log_gdp*). That is, female literacy tends to be missing for the more developed countries!

Figure 2.3   Profile of means for the female literacy pattern

In Figure 2.4, relationships between pairs of variables are displayed in scatterplots. The missing data pattern for female literacy is highlighted in the left plot; that for calories, in the right plot.

Figure 2.4   Patterns for female literacy and calories

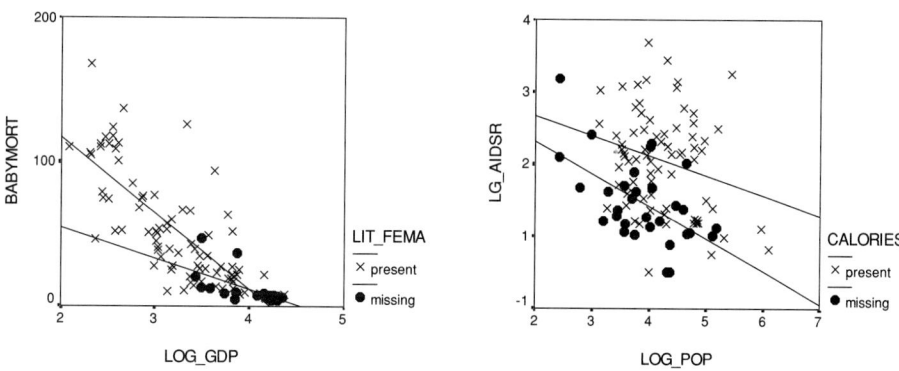

Within the scatterplot, regression lines are drawn for each group. In the plot on the right, the pattern for calories separates the values of population and AIDS rate into two groups. The top regression line fits the countries where *calories* is present; the lower line, where *calories* is missing.

Crosstabulating patterns against categorical variables. The user can specify categorical variables against which the predominant pattern variables are crosstabulated. Here, we request tables for *region2* and *religion*.

REGION2

			Total	Europe	East Europe	Pacific/Asia	Africa	Middle East	Latn America	Missing SysMis
CALORIES	Present	Count	75	14	3	13	16	8	19	2
		Percent	68.8	82.4	21.4	68.4	84.2	47.1	90.5	100.0
	Missing	% SysMis	31.2	17.6	78.6	31.6	15.8	52.9	9.5	.0
LIT_MALE	Present	Count	85	4	9	16	18	16	21	1
		Percent	78.0	23.5	64.3	84.2	94.7	94.1	100.0	50.0
	Missing	% SysMis	22.0	76.5	35.7	15.8	5.3	5.9	.0	50.0
LIT_FEMA	Present	Count	85	4	9	16	18	16	21	1
		Percent	78.0	23.5	64.3	84.2	94.7	94.1	100.0	50.0
	Missing	% SysMis	22.0	76.5	35.7	15.8	5.3	5.9	.0	50.0

The count of values present for each category is given in the first row of each pattern variable (for example, of the 75 *calories* values that are present, 14 are in the Europe category). The percentage each count is of the total sample size is given in the next row (75 is 68.8% of 109 countries and 14 is 82.4% of the 17 European countries). The percentage missing for individual categories can be contrasted with the overall percentage missing reported in the *Total* column. Overall, 31.2% of the values of *calories* are missing, but the variable is missing for more than three-fourths (78.6%) of the Eastern European and half (52.9%) the Middle Eastern countries. The information for *lit_fema* is missing for three-fourths (76.5%) of the European countries. Across the *region2* categories, the Latin American countries have many fewer missing values.

			Total	Hindu	Jewish	Tribal	Muslim	Taoist	Animist	Catholic	Protstnt	Buddhist	Orthodox	Missing
CALORIES	Present	Count	75	1	0	1	15	2	4	35	11	4	2	0
		Percent	68.8	100.0	.0	100.0	55.6	100.0	100.0	85.4	68.8	57.1	25.0	.0
	Missing	% SysMis	31.2	.0	100.0	.0	44.4	.0	.0	14.6	31.3	42.9	75.0	100.0
LIT_MALE	Present	Count	85	1	1	1	25	2	4	32	8	5	6	0
		Percent	78.0	100.0	100.0	100.0	92.6	100.0	100.0	78.0	50.0	71.4	75.0	.0
	Missing	% SysMis	22.0	.0	.0	.0	7.4	.0	.0	22.0	50.0	28.6	25.0	100.0
LIT_FEMA	Present	Count	85	1	1	1	25	2	4	32	8	5	6	0
		Percent	78.0	100.0	100.0	100.0	92.6	100.0	100.0	78.0	50.0	71.4	75.0	.0
	Missing	% SysMis	22.0	.0	.0	.0	7.4	.0	.0	22.0	50.0	28.6	25.0	100.0

RELIGION

In the crosstabulation of patterns with *religion,* values of *lit_fema* and *lit_male* are missing for half of the countries (50%) that are predominantly Protestant—this is considerably more than that found for the other religions. Overall, only 22% of the values are missing.

Example 3:
Patterns in a Large Survey

Sometimes data are missing by design. The General Social Survey has been administered yearly since 1972. Each survey is an independently drawn sample of people over 18 years of age living in non-institutional arrangements. Each year, permanent items are included; but starting in 1988, a subset of items appears on two-thirds of the cases each year. Actually, there are three subsets of items, and each is asked of a different two-thirds subsample of the respondents. The codebook we received does not identify which items are in what subset, nor which respondents received them. In the crosstabulation of pattern variables for questions about gun laws, marijuana, and euthanasia (Table 2.1), we show that there are no cases that are complete across the three items. We used the Tabulated Pattern display in this example to identify the three subsets of items and then selected one item from each as table factors.

The responses for many survey items have dichotomous or ordered categories. For a first look at the survey, we treat these items as quantitative variables. If we were to pursue these data further, we would identify them as categorical variables for some purposes—for example, when crosstabulating patterns against categorical variables.

Because the *gss93mva* data file is considerably larger than that used in the first examples, the allocation for memory needs to be increased. To do this, from the menus choose:

Edit
  Options...

Special Workspace Memory Limit
  1250 K Bytes

(When you are finished working with the large file, reset the memory limit to 512K.)

Now you are ready to run the Missing Value procedure. To produce the following output, from the menus choose:

Statistics
  Missing Value Analysis

▶ Quantitative Variables: childs, age, educ, degree, padeg, madeg, sex, income91, rincom91, size, vote92, polviews, natenvir, natheal, natcity, natcrime, natdrug, nateduc, cappun, gunlaw, grass, life, sathobby, drink, fework, pillok, spanking, letdie1, news, tvhours, bigband, blugrass, country, blues, musicals, classicl, folk, jazz, opera, rap, hvymetal, attsprts, visitart, tvshows, tvnews, tvpbs, partners, sexfreq, sei

▶ Categorical Variables: wrkstat, marital, birthmo, zodiac, race, region, xnorcsiz, partyid, relig, aged, sexeduc, dwelown

  Estimation
  ☑ Pairwise

Patterns...
  Display
  ☑ Tabulated cases, grouped by missing value patterns

Descriptives...
  Indicator Variable Statistics
  ☑ Percent mismatch
  ☑ *t* tests with groups formed by indicator variables
  ☑ Crosstabulations of categorical and indicator variables
  Omit variables missing less than **20%** of cases

Note: You may wish to run *Tabulated cases* under *Patterns* separately because it could take a good bit of time.

Univariate statistics. For this display, we pasted the syntax and rearranged the variables list to reflect the order in the file. This univariate panel provides an overview of the extent of incomplete data. Because the panel is long, we split it into two sections. Nine variables are missing roughly half (from 49.5% to 56.7%) of their values, starting with *region* and ending with the *nation* items (whether the nation is spending *too little, about right,* or *too much* for the environment, health, etc.). Fourteen variables are missing about one-third of their values. Among these must be the items given to only two-thirds of the sample. In the Tabulated Patterns display, we will try to distinguish the three subsets of items.

For this larger data set, the criterion for identifying extreme values is now the mean plus or minus two standard deviations. If the product of the number of variables times the number of cases times the base 10 log of the number of cases is less than 150,000, the robust measure defined in Example 1 is used; otherwise, the rule here is used.

For *age*, there are 41 "high" extreme values. If you run the Frequencies procedure with a histogram, you will find there are 41 respondents older than 81 years (46.23 + 2*17.42) and that the age distribution is right-skewed. Watch out for outliers, however, when the data are coarse (each variable has only a few unique values). For example, the responses for *degree*, *padeg*, and *madeg* (father's and mother's degrees) have five codes with 3 meaning bachelor's degree and 4, a graduate degree. From the Frequencies procedure, you will find that the 113 high values for *degree* are the count of graduate degrees and the 125 high values for *madeg* include 96 bachelor degrees and 29 graduate degrees.

Univariate Statistics

	N	Mean	Std. Deviation	Missing		No. of Extremes	
				Count	Percent	Low	High
WRKSTAT	1500			0	.0		
MARITAL	1499			1	.1		
CHILDS	1495	1.85	1.68	5	.3	0	42
AGE	1495	46.23	17.42	5	.3	0	41
BIRTHMO	1487			13	.9		
ZODIAC	1487			13	.9		
EDUC	1496	13.04	3.07	4	.3	40	25
DEGREE	1496	1.41	1.18	4	.3	0	113
PADEG	1207	.93	1.19	293	19.5	0	72
MADEG	1352	.84	.94	148	9.9	0	125
SEX	1500	1.57	.49	0	.0	0	0
RACE	1500			0	.0		
INCOME91	1434	14.68	5.46	66	4.4	59	0
RINCOM91	994	12.80	5.62	506	33.7	26	0
REGION	757			743	49.5		
XNORCSIZ	757			743	49.5		
SIZE	757	360.63	1225.20	743	49.5	0	42
PARTYID	1491			9	.6		
VOTE92	1492	1.34	.54	8	.5	0	40
POLVIEWS	1443	4.17	1.36	57	3.8	30	41
NATENVIR	710	1.51	.67	790	52.7	0	69
NATHEAL	720	1.33	.61	780	52.0	0	54
NATCITY	650	1.52	.72	850	56.7	0	88
NATCRIME	717	1.32	.57	783	52.2	0	38
NATDRUG	707	1.44	.64	793	52.9	0	57
NATEDUC	724	1.37	.59	776	51.7	0	43
CAPPUN	1388	1.23	.42	112	7.5	0	0
GUNLAW	984	1.18	.38	516	34.4	0	173
GRASS	930	1.77	.42	570	38.0	0	0
RELIG	1492			8	.5		
LIFE	997	2.41	.61	503	33.5	65	0
SATHOBBY	970	2.65	1.52	530	35.3	0	63
AGED	967			533	35.5		
DRINK	981	1.31	.46	519	34.6	0	0

	N	Mean	Std. Deviation	Missing		No. of Extremes[a]	
				Count	Percent	Low	High
FEWORK	992	1.20	.40	508	33.9	0	198
PILLOK	974	2.34	1.07	526	35.1	0	0
SEXEDUC	984			516	34.4		
SPANKING	997	2.11	.83	503	33.5	0	67
LETDIE1	956	1.32	.47	544	36.3	0	0
NEWS	1010	2.02	1.21	490	32.7	0	58
TVHOURS	1489	2.90	2.24	11	.7	0	63
BIGBAND	1337	2.45	1.09	163	10.9	0	52
BLUGRASS	1335	2.66	1.02	165	11.0	0	56
COUNTRY	1468	2.32	1.09	32	2.1	0	63
BLUES	1434	2.51	1.03	66	4.4	0	57
MUSICALS	1412	2.60	1.09	88	5.9	0	64
CLASSICL	1425	2.66	1.22	75	5.0	0	0
FOLK	1414	2.76	1.04	86	5.7	0	85
JAZZ	1451	2.62	1.11	49	3.3	0	66
OPERA	1410	3.49	1.13	90	6.0	66	0
RAP	1431	3.93	1.12	69	4.6	41	0
HVYMETAL	1423	4.13	1.11	77	5.1	45	0
ATTSPRTS	1489	1.47	.50	11	.7	0	0
VISITART	1488	1.60	.49	12	.8	0	0
TVSHOWS	1490	2.52	1.19	10	.7	0	94
TVNEWS	1492	1.60	1.00	8	.5	0	118
TVPBS	1486	2.71	1.25	14	.9	0	0
PARTNERS	1367	1.02	.93	133	8.9	0	83
SEXFREQ	1330	2.88	1.98	170	11.3	0	0
DWELOWN	1012			488	32.5		
SEI	1419	47.206	18.760	81	5.4	0	56

a. Number of cases outside the range (Mean-2*SD, Mean+2*SD).

Common patterns of missing data. In the Tabulated Patterns display, the order of the variables results from simultaneously sorting the rows and columns of the data matrix by their missing value patterns. If 15 (1% of 1500) or fewer cases have the same pattern, they are not tabulated here. The last 22 variables are the ones identified in the Univariate Statistics panel as having one-third or more of their values missing.

Tabulated Patterns

# of cases	WRKSTAT	SEX	RACE	MARITAL	AGE	CHILDS	DEGREE	EDUC	VOTE92	PARTYID	RELIG	TVNEWS	TVSHOWS	TVPBS	ATTSPRTS	VISITART	TVHOURS	BIRTHMO	ZODIAC	COUNTRY	JAZZ	BLUES	RAP	HVYMETAL	FOLK	MUSICALS	OPERA	POLVIEWS	INCOME91	SEI	CAPPUN	PARTNERS	SEXFREQ	MADEG	BIGBANG	BLUGRASS	PADEG	RINCOM91	SEXEDUC	DWELOWN	NEWS	SPANKING	FEWORK	PILLOK	LETDIE1	NATCRIME	NATEDUC	NATHEAL	NATENVIR	NATDRUG	NATCITY	GRASS	DRINK	SATHOBBY	AGED	REGION	XNORCSIZ	SIZE	LIFE	GUNLAW	Complete if...
0																																																														0
34																																																											X	X	34
33																																																									X	X	X	X	67
44																																															X	X	X	X	X	X	X						X	X	X	X	X	137
33																																															X	X	X	X	X	X	X								X	X	62
23																																																						X	X	X	X						23
26																																																						X	X	X	X	X	X	X				49
44																																															X	X	X	X	X	X	X	X	X	X	X	X	X				126
47																																															X	X	X	X	X	X	X	X	X	X							60
36																																												X	X	X	X	X	X	X	X	X	X	X	X								63
39																																												X	X	X	X	X	X	X	X	X	X	X					X	X	X			129
35																																												X	X	X	X	X	X	X												35
29																																												X	X	X	X	X	X	X								X	X	X			64

To collect cases that may have a few other values missing in addition to those here, let's digress and rerun the same analysis using only the last 22 variables.

Tabulated Patterns

Number of Cases	DWELOWN	NEWS	SPANKING	FEWORK	SEXEDUC	PILLOK	LETDIE1	NATCRIME	NATEDUC	NATHEAL	NATENVIR	NATDRUG	NATCITY	GRASS	DRINK	SATHOBBY	AGED	REGION	XNORCSIZ	SIZE	LIFE	GUNLAW	Complete if...
0																								0
82																						X	X	82
90				X	X	X	X	X	X													X	X	157
108				X	X	X	X	X	X									X	X	X	X	X	X	341
95																		X	X	X	X	X	X	177
124				X	X	X	X	X	X	X	X	X	X											184
86														X	X	X	X							86
86														X	X	X	X	X	X	X				172
109				X	X	X	X	X	X	X	X	X	X	X	X	X	X	X	X				363
89	X	X	X	X	X	X	X																	89
87	X	X	X	X	X	X	X									X	X	X						176
102	X	X	X	X	X	X	X	X	X	X	X	X				X	X	X						363
123	X	X	X	X	X	X	X	X	X	X	X	X												192

In the Tabulated Patterns display restricted to 22 variables, we highlight three sets of variables that are missing independently of the others: 1) *dwelown, news, spanking, fework, sexeduc, pillok, letdie1*; 2) *grass, drink, sathobby, aged*; and 3) *life* and *gunlaw*.

Each of these sets is sometimes missing with the *"nat"* set of six variables or with the set that includes *region*, but not simultaneously with another of these sets. We will look for these sets in the Percent Mismatch table below.

Pairwise mismatched patterns. Only variables that have 20% or more of their values missing are included in the Percent Mismatch table. Diagonal entries are the percentage missing for each variable individually; off the diagonal, the entries are the percentage with one member of the pair missing and the other present. We highlight pairs combining the three sets of variables identified above:

Percent Mismatch of Indicator Variables.

	RINCOM91	SEXEDUC	DWELOWN	NEWS	SPANKING	FEWORK	PILLOK	LETDIE1	NATCRIME	NATEDUC	NATHEAL	NATENVIR	NATDRUG	NATCITY	GRASS	DRINK	SATHOBBY	AGED	XNORCSIZ	REGION	SIZE	LIFE	GUNLAW
RINCOM91	34																						
SEXEDUC	43	34																					
DWELOWN	44	2	33																				
NEWS	44	2	0	33																			
SPANKING	44	3	1	1	34																		
FEWORK	44	3	2	2	2	34																	
PILLOK	45	4	3	3	3	4	35																
LETDIE1	45	5	4	4	5	5	5	36															
NATCRIME	51	50	51	50	50	50	50	50	52														
NATEDUC	51	50	51	51	50	51	51	50	2	52													
NATHEAL	51	50	51	51	50	51	51	50	2	2	52												
NATENVIR	51	50	51	51	50	50	50	50	3	2	3	53											
NATDRUG	50	51	52	51	51	51	51	51	3	3	3	3	53										
NATCITY	52	51	52	51	51	51	51	50	6	7	7	7	7	57									
GRASS	46	66	66	66	66	67	65	66	50	49	50	49	50	50	38								
DRINK	45	67	67	67	67	68	66	67	50	49	50	50	50	51	3	35							
SATHOBBY	45	66	67	67	66	67	66	66	50	49	49	49	50	50	4	1	35						
AGED	45	66	66	67	66	67	66	66	50	49	50	50	50	51	4	1	1	36					
XNORCSIZ	50	51	51	51	51	51	51	51	51	51	51	51	51	51	51	51	51	51	50				
REGION	50	51	51	51	51	51	51	51	51	51	51	51	51	51	51	51	51	51	0	50			
SIZE	50	51	51	51	51	51	51	51	51	51	51	51	51	51	51	51	51	51	0	0	50		
LIFE	45	65	66	65	65	66	65	51	52	51	52	51	54	68	68	68	68	48	48	48	34		
GUNLAW	44	65	65	65	66	65	66	65	51	51	51	52	51	53	68	68	68	67	48	48	48	2	34

When a variable is paired with a variable from one of the other sets, roughly two-thirds of the cases are mismatched. For example, 34% of the *gunlaw* values and 38% of the *grass* values are missing—yet, when paired together, 68% of the cases have either *gunlaw* or *grass* missing, but not both.

Identifying nonrandom patterns with t tests. As in Example 2, we pivot the results of the separate variances two-sample *t* tests into layers displaying the panel of *t* statistics alone (drag the statistics icon into the Layer tray). In addition, *we transpose the rows and columns* for a better fit on the page. Thus, the pattern variables that form groups for the two-sample *t* tests are columns and the measures tested, the rows. The panel shown here has already been pivoted into layers and transposed.

Separate Variance t Tests

	RINCOM91	SIZE	NATENVIR	NATHEAL	NATCITY	NATCRIME	NATDRUG	NATEDUC	GUNLAW	GRASS	LIFE	SATHOBBY	DRINK	FEWORK	PILLOK	SPANKING	LETDIE1	NEWS
CHILDS	-6.7	1.3	-2.2	-1.7	-2.3	-1.8	-1.4	-1.5	-1.4	-.8	-1.4	-.8	-.5	1.2	1.6	2.2	1.6	1.8
AGE	-15.9	3.5	-1.8	-1.7	-2.9	-1.9	-2.1	-1.9	-.4	-.1	.3	-1.0	-.3	-.5	-1.3	-.2	-.4	-.2
EDUC	12.0	-.4	1.3	.7	2.2	.6	.5	.8	.5	.6	.1	1.2	.8	-.2	-.1	-.9	-.1	-.8
DEGREE	11.4	.0	1.0	.8	2.2	.7	.6	.7	.5	.8	-.1	.8	.5	.2	-.1	-.6	.5	-.3
PADEG	6.2	-2.9	3.0	3.0	3.8	3.2	3.1	3.0	.6	-1.0	.4	-.9	-1.0	1.2	1.3	1.1	1.3	.9
MADEG	6.3	-1.8	2.0	1.8	2.5	1.9	1.7	1.6	1.2	-.9	.9	-.2	-.4	.3	.0	-.1	.5	-.1
SEX	-6.6	-.3	-2.4	-1.7	-2.9	-2.3	-2.0	-1.9	-.5	1.8	.0	1.9	2.2	-2.8	-2.1	-2.1	-2.7	-2.4
INCOME91	12.6	.3	-.3	-.5	.6	-.2	-.4	-.4	-.2	1.0	.0	1.3	1.3	-.8	-1.5	-1.3	-.9	-1.3
RINCOM91	.	1.5	1.2	.9	2.0	1.2	.7	1.3	.9	.3	1.1	-.3	.0	-1.1	-1.9	-1.6	-.5	-1.3
SIZE	1.8	.	.6	.9	1.1	1.0	.8	.4	.6	-.6	.5	.2	.1	-.6	-.4	-.3	-1.0	-.4
VOTE92	-2.4	-1.0	-.1	.2	-1.0	-.6	.4	.1	-2.1	-.3	-1.8	.3	.5	.0	.3	.5	.6	1.0
POLVIEWS	-3.3	1.5	.0	.1	.3	.3	.1	.2	-.7	1.3	-.6	.9	.9	-.7	.0	-.5	-.8	-.4
NATENVIR	-4.4	2.2	.	-.6	-1.4	.5	-2.1	-.9	-.7	-.8	-.8	-1.1	-.8	1.2	.6	1.5	1.6	1.6
NATHEAL	-2.8	.5	-.2	.	.3	3.2	-.9	-.6	.5	-.2	.2	-.7	-.4	.1	-.5	-.1	.3	-.1
NATCITY	-1.0	1.1	-.5	-.6	.	-.1	-1.2	-.7	-.8	.7	-.6	.9	1.0	-.6	-.5	-.3	-.5	-.4
NATCRIME	-.2	-.8	-.6	-1.2	.4	.	-.7	-.1	2.6	-.2	2.1	-.8	-.7	-1.6	-1.5	-1.5	-1.2	-1.8
NATDRUG	1.6	.2	-1.4	-1.4	-.9	-.7	.	-.1	.5	.6	.5	.5	.6	-1.6	-1.8	-1.2	-1.2	-1.2
NATEDUC	-3.0	-.3	-2.1	-.4	-.9	-1.3	-2.7	.	-.1	-2.0	-.1	-1.8	-1.7	1.7	1.2	2.1	1.6	1.7
CAPPUN	-1.4	-1.3	-.4	-.7	-.8	-.8	-.7	-.6	-1.4	1.1	-1.4	1.0	.9	.4	1.0	.7	.7	.6
GUNLAW	-1.1	1.2	3.2	3.0	2.6	3.2	3.3	3.1	.	1.0	-.1	.3	.6	-.6	.2	-.5	-.3	-.5
GRASS	-2.7	3.2	-1.1	-1.4	-1.6	-1.4	-1.2	-1.2	-.2	.	.0	-16.5	.	-.1	-.1	.2	-.4	.0
LIFE	4.8	.2	-.9	-1.1	-.2	-.8	-1.3	-.7	2.8	-.3	.	-.3	-.6	.7	.5	.6	.5	.6
SATHOBBY	-3.3	-2.5	1.1	1.3	1.8	1.0	1.7	1.1	-3.6	.2	-3.8	.	.	3.0	3.5	3.4	2.8	3.7
DRINK	-6.4	.3	-1.6	-1.7	-2.0	-1.4	-1.8	-1.5	.0	1.7	.4	-4.2	.	-1.1	-.5	-.3	-1.0	-.7
FEWORK	-4.7	1.8	-1.8	-1.8	-1.5	-1.8	-1.9	-1.9	.4	-.3	.4	-.7	-.3	.	-.3	.1	.5	-.2
PILLOK	-3.8	1.5	-2.4	-2.2	-2.1	-2.6	-2.5	-2.6	-1.1	1.5	-.9	.6	1.0	-.2	.	1.0	-2.1	-2.3
SPANKING	1.8	1.1	1.4	1.2	1.8	1.1	1.2	1.4	-1.6	1.8	-1.8	1.9	1.8	1.8	-.6	.	-1.0	4.0
LETDIE1	-3.8	2.5	-.2	-.2	-.4	-.1	-.2	-.5	2.0	-1.9	2.5	-2.4	-2.4	-.9	.4	-.3	.	.0
NEWS	-1.2	1.8	-.3	.0	-1.3	.0	.6	.0	-1.0	.8	-.9	.7	.9	-1.8	-.2	.7	.0	.
TVHOURS	-9.3	-.4	-.6	-.9	-1.2	-.7	-.5	-1.2	-.3	-.8	-.3	-.5	-.6	.8	.8	.9	.6	.9
BIGBAND	2.6	-2.0	.6	.6	.3	.5	1.0	1.1	-.7	.1	-.8	-.2	-.1	1.0	.5	1.1	1.3	.9
BLUGRASS	.4	-2.1	-1.2	-.5	-.4	-1.1	-1.1	-1.0	-.3	1.0	-.6	1.3	1.5	-1.5	-1.0	-.9	-.7	-1.0
COUNTRY	.5	-2.1	1.1	1.4	1.1	.8	.9	.9	-.1	.3	-.2	.7	.8	-1.0	-.7	-.6	.2	-.7
BLUES	-3.6	.9	-1.8	-1.9	-2.4	-2.0	-1.8	-1.9	-1.5	-.9	-1.3	-.8	-.7	1.6	1.2	2.3	1.8	2.1
MUSICALS	-.4	-.6	.0	-.4	-.1	.0	.0	-.1	-1.0	-.4	-1.0	-.3	-.2	.9	1.1	1.6	1.9	1.3
CLASSICL	-2.6	-.5	-.4	-.7	-.6	-.2	-.3	-.4	-.7	-.7	-.6	-.8	-.9	1.0	.7	1.8	1.5	1.5
FOLK	-.6	-2.3	-.6	-.3	.5	-.4	-.4	-.3	-1.9	.5	-1.8	1.5	1.4	-.1	-.2	.4	.5	.4
JAZZ	-4.4	.6	-1.4	-1.2	-2.0	-1.6	-1.6	-1.0	-.2	-1.0	.2	-1.8	-1.5	1.4	1.0	1.7	1.4	1.4
OPERA	1.5	-.2	1.0	.9	1.0	1.2	1.2	1.0	-.7	-.7	-.6	-1.1	-1.0	1.7	1.2	2.0	2.0	1.7
RAP	-.8	-1.2	-.2	-.5	-.5	-.1	-.4	-.7	-.4	1.1	-.7	.9	.8	.0	-.4	-.2	.1	-.1
HVYMETAL	-5.2	1.8	.6	.7	.5	.9	.8	.6	1.7	.0	1.5	-.4	-.3	-1.4	-1.7	-1.2	-1.7	-1.2
ATTSPRTS	-9.5	.0	-1.3	-1.1	-2.4	-1.4	-.9	-1.1	.1	.0	.4	-.5	.0	-1.2	-1.2	-.8	-.9	-.8
VISITART	-6.3	1.4	-1.2	-1.4	-2.3	-1.1	-.9	-1.4	-1.0	-.3	-.7	-.4	-.3	.4	.4	1.3	.5	.9
TVSHOWS	.4	.9	-.3	.4	-.1	-.2	-.2	.4	2.2	.8	2.6	.2	.7	-3.7	-3.3	-3.5	-3.1	-3.5
TVNEWS	3.0	.1	-1.2	-.3	-.7	-.9	-.3	-.4	-.4	-.6	.1	-.5	-.4	-.3	.3	.2	.0	.2
TVPBS	1.1	.8	.2	.0	-.7	.6	.6	.6	.1	-.4	.5	-.6	-.4	-.7	-.5	.1	-.3	-.2
PARTNERS	8.9	-3.0	2.1	1.7	1.9	1.5	1.9	1.9	-.5	.1	-.5	.3	.0	.8	1.1	.9	1.2	.7
SEXFREQ	11.8	-2.2	1.9	1.1	2.1	1.6	1.7	1.7	-2.5	1.0	-2.7	2.1	1.6	1.5	1.9	1.4	1.7	1.2
SEI	6.8	.9	-.3	-.5	.5	-.5	-.9	-.5	.8	.8	.5	.7	.6	-.5	-1.1	-.8	.1	-.6

The most extreme *t* value is −16.5 for the *grass* item split into groups based on the pattern of *sathobby* (the respondents satisfaction with nonworking activities, or hobbies). In the full table with all layers displayed (not shown), we find that only eight cases have values of *sathobby* missing when *grass* is present; so we do not pursue this relation further. These variables belong to the same subset identified above and, if missing, tend to be missing together.

The tests based on the pattern of *rincom91* (respondent's income on the 1991–93 surveys) are more interesting. The *t* statistics are sizable for more than 20 items, including *income91* (total family income from all sources). See Figure 2.5 for boxplots of four of these distributions: age, education, income, and political views grouped by the pattern of *rincom91*.

Figure 2.5 Boxplots of age, education, family income, and political views grouped by the pattern of rincom91

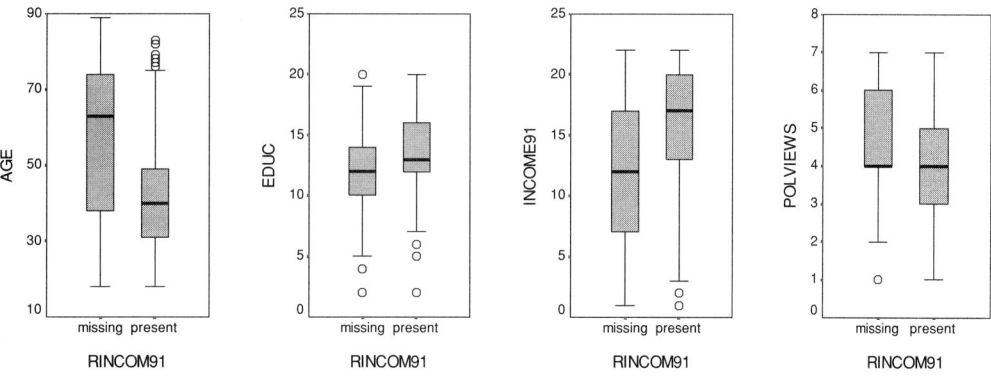

The 34% of the respondents for whom *rincom91* is missing tend to be older, have less education, have a lower total family income, and are more conservative (low values of *polviews* indicate liberal political views) than the 66% who do report *rincom91*. If *rincom91* is crosstabulated against *wrkstat* (this variable's eight codes define job status), more than 80% of the respondents with missing *rincom91* values report they are *retired*, *in school*, or *a homemaker*.

A Look at Multiple Missing Value Codes

When the pattern variable has a few distinct categories, it is easy to use SPSS's boxplots to compare distributions of a measure within valid categories against those for one or more types of missing values.

In the previous two-sample *t* tests, each pattern variable had to have at least 20% of its values missing. Lowering this cutoff to the default value of 5%, would add the big band and heavy metal music items as pattern variables. The two-sample *t* statistics for testing differences in average age for the groups formed by the two patterns are respectively, –6.95 and 6.24. The average age for the 163 subjects with missing big band values is 37.9 years; the average age for the 77 people with missing heavy metal values is 60.4 years. The variables *bigband* and *hvymetal* have two missing value codes: 8 for *don't know much* and 9 for *no answer*. Figure 2.6 and Figure 2.7 examine the age distribution within each of these missing value categories.

Figure 2.6 People who "don't know much" about big band music are younger

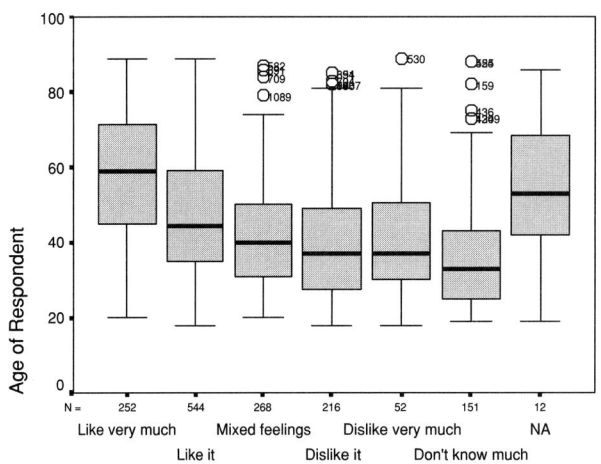

In both Figure 2.6 and Figure 2.7, the age distribution of the few subjects with *no answer* (*NA*) differs from that for the *don't know much* group.

Figure 2.7 People who "don't know much" about heavy metal music are older

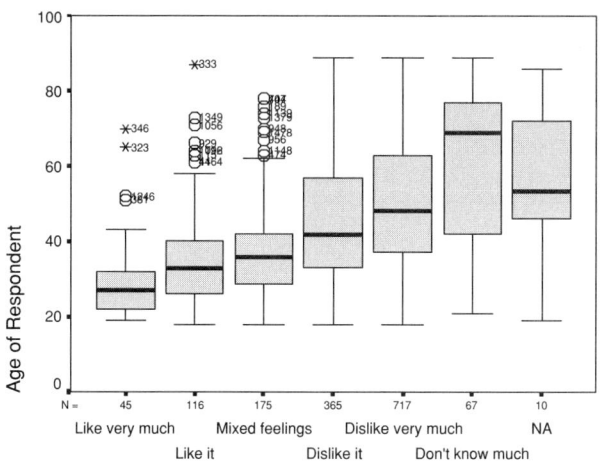

The Missing Value feature for crosstabulating pattern variables against categorical variables is useful for studying relationships among categories with different missing value codes. Here we define *polviews* (Where do you place yourself on a scale of political views ranging from *Extremely liberal* to *Extremely conservative*?) as a categorical variable and show how the categories relate to selected patterns. To produce the output, we specify:

Descriptives...

Indicator Variable Statistics
☑ Crosstabulation of categorical and indicator variables
  Omit variables missing less than **5%** of cases

The pattern variables include *cappun* (Do you favor or oppose the death penalty for persons convicted of murder?), *letdie1* (If a person is in an advanced stage of a terminal illness, should doctors be allowed to end the patient's life if he and his family request it?), and the respondent's preference for big band and heavy metal music.

Overall, 7.1% of the respondents answered that they *don't know* about capital punishment, but this response is spread unevenly across the seven categories. Almost 10% of the *moderates* say they don't know, while the percentages for those in the three *liberal* and three *conservative* categories are collectively closer to 5%. Among those who *don't know* (*DK*) where they belong on the political scale, 18.8% also don't know where they stand regarding capital punishment (this is considerably higher than the overall 7.1% rate).

POLVIEWS

			Total	Extremely liberal	Liberal	Slightly liberal	Moderate	Slightly conservative	Conservative	Extremely conservative	Missing DK	Missing NA
CAPPUN	Present	Count	1388	29	152	184	473	238	228	39	39	6
		Percent	92.5	96.7	93.3	95.3	89.8	96.0	94.6	95.1	81.3	66.7
	Missing	% DK	7.1	3.3	5.5	4.7	9.9	3.6	5.4	4.9	18.8	22.2
		% NA	.4	.0	1.2	.0	.4	.4	.0	.0	.0	11.1
LETDIE1	Present	Count	956	22	103	125	344	161	148	26	25	2
		Percent	63.7	73.3	63.2	64.8	65.3	64.9	61.4	63.4	52.1	22.2
	Missing	% NAP	32.3	26.7	35.6	32.1	29.4	33.1	34.4	31.7	41.7	44.4
		% DK	3.7	.0	1.2	3.1	4.9	2.0	4.1	4.9	6.3	11.1
		% NA	.3	.0	.0	.0	.4	.0	.0	.0	.0	22.2
BIGBAND	Present	Count	1337	25	145	167	465	227	224	39	40	5
		Percent	89.1	83.3	89.0	86.5	88.2	91.5	92.9	95.1	83.3	55.6
	Missing	% Don't know	10.1	13.3	10.4	13.0	11.6	8.1	6.2	4.9	12.5	11.1
		% NA	.8	3.3	.6	.5	.2	.4	.8	.0	4.2	33.3
HVYMETAL	Present	Count	1423	28	160	182	508	238	229	38	35	5
		Percent	94.9	93.3	98.2	94.3	96.4	96.0	95.0	92.7	72.9	55.6
	Missing	% Don't know	4.5	3.3	1.2	5.2	3.6	3.6	4.6	7.3	22.9	11.1
		% NA	.7	3.3	.6	.5	.0	.4	.4	.0	4.2	33.3

In the *Extremely liberal* and *Liberal* categories, there are fewer folks (0.0% and 1.2%) who are uncertain (*DK*) about euthanasia than there are in the conservative groups (4.1% and 4.9%). With respect to big band music, there is a linear decrease in the percentage answering *don't know* across the seven political categories (13.3%, 10.4%, 13%, 11.6%, 8.1%, 6.2%, and 4.9%)—conservatives are less apt to report that they don't know about big band music. The opposite is true for heavy metal music.

Example 4:
Estimating Means, Standard Deviations, Covariances, and Correlations

In the Missing Value procedure, the user can choose to estimate means, standard deviations, covariances, and correlations using listwise (complete cases only), pairwise, EM, and/or regression methods. The Missing Value procedure also provides all values estimates of means and standard deviations plus several options for the EM and regression methods. When more than one method is requested, the estimates of the means are displayed in one summary panel, and the estimates of the standard deviations are displayed in another (see p. 46).

Over the years, many software users approached the missing data problem by using a pairwise complete method to compute a covariance or correlation matrix and then using this matrix as input for, say, a factor analysis. However, such a matrix may have eigenvalues less than 0, and some correlations may be computed from substantially different subsets of the cases. Other analysts use EM (expectation-maximization) or regression methods to estimate statistics or to impute data (estimate replacement values). Simulation studies indicate that pairwise estimates are often more distorted than estimates obtained via the EM method. In most algorithms, they are simply the first iteration of the EM method. A few analysts use multiple imputation, a computationally complex method that is not commonly available. For a variation of this method, see "Multiple Imputation" on p. 59.

Listwise method. This method uses complete cases only. That is, if among the variables you select as quantitative or categorical there are one or more values missing, the case is omitted from the computations.

Pairwise method. Estimates are computed separately for each pair of variables, using all cases that have both values.

EM method. For the EM procedure, a distribution is assumed for the partially missing data, and inferences are based on the likelihood under that distribution. Each iteration consists of an E step and an M step. The E step finds the conditional expectation of the "missing" data, given the observed values and current estimates of the parameters. These expectations are then substituted for the "missing" data. In the M step, maximum likelihood estimates of the parameters are computed as though the missing data had been filled in. "Missing" is enclosed in quotation marks because the missing values are not being directly filled, but, rather, functions of them are used in the log-likelihood.

By default for the EM method, the Missing Value procedure assumes that the data follow a normal distribution. If you know that the tails of the distributions are longer than those of a normal distribution, you can request that a t distribution with n degrees of freedom be used in constructing the likelihood function (n is specified by the user). A second option also provides a distribution with longer tails. You specify the ratio of standard deviations of a mixed normal distribution and the mixture proportion of the two distributions. This assumes that only the standard deviations of the distributions differ, not the means.

The default is to use all variables on the Quantitative Variables list in the Missing Value Analysis dialog box for estimation. However, in the Variables for EM and Regression dialog box, you can specify that specific variables are predictor variables or predicted variables. Of course, a given variable can be in both lists, but there are situations in which you might want to restrict the use of a variable. For example, some analysts are uncomfortable estimating values of outcome variables. Or, if, for each subject, you have a set of items that are nurses' ratings and another set that are doctors' ratings, you may want to make one run using the nurses' items to estimate missing nurses' items and another run for estimates of the doctors' items.

Regression method. To estimate means, standard deviations, covariances, or correlations, the regression method computes multiple linear regression estimates and has options for augmenting the estimates with random components. To each predicted value, the Missing Value procedure can add a residual from a randomly selected complete case, a random normal (0, RMS) deviate, or a random deviate (scaled by the square root of the residual mean square) from the t distribution with n degrees of freedom (the user can specify n or use the default value of 5). To add nothing, select *None* in the Regression dialog box. The default is to add a randomly selected residual. However, if the number of complete cases is less than half the total sample size, a normal deviate is added.

All quantitative variables specified as predictors are available as candidates for estimation (see the discussion of the EM method about use of variables). In addition, since in multiple regression the use of a large subset of independent variables can produce poorer predicted values than a smaller subset, a variable must achieve an F-to-enter limit of 4.0 to be used. This limit can be changed with syntax.

Assumptions. If data are **missing completely at random** (called MCAR by Little and Rubin), complete cases, pairwise, EM, and regression methods give consistent and unbiased estimates of correlations and covariances. The pairwise, EM, and regression methods may still provide good estimates if the data are *conditionally* **missing at random** (MAR). For example, in a study of education and income, the subjects with low education may have more missing income values. If education is MCAR and if, for a given level of education, income is MCAR, pairwise, EM, and regression methods may still give good estimates. In other words, for MAR, the probability that income is recorded depends on the subject's level of education, so the probability may vary by education but not by income *within that level of education.* Besides MCAR and MAR patterns, the probability that income is present could vary by the value of income within each level of education (for example, people with high incomes don't report them). The last situation is not an unusual pattern for real-world applications, but, alas, current methods are not appropriate.

If the data are MAR and the assumption that the distributions are normal, mixed normal, or t with specific degrees of freedom is met, the EM methods yield maximum likelihood estimates of means, standard deviations, covariances, and correlations. Be sure to check the data for outliers and to determine whether symmetrizing transformations are required.

Roderick J. A. Little's chi-square statistic for testing whether values are missing completely at random is printed with EM matrices (see p. 48). The separate variances two-sample t tests introduced in Example 2 are also useful for identifying departures from randomness. However, be aware that while a sizable t statistic does indicate a departure from randomness, a small t may be no confirmation that values are missing randomly (see the discussion with Figure 2.9). Sadly, there is no magic test for MAR.

In this example, we continue to use the *world95m* data used in Example 2, now requesting estimates of statistics. Even though we established that values are nonrandomly missing, we request listwise (complete cases) estimates so that they can be compared later with estimates obtained by the pairwise, EM, and regression methods. Estimates of means and standard deviations obtained by using all available data for each variable are displayed automatically.

To produce this output, from the menus choose:

Statistics
 Missing Value Analysis...

▶ Quantitative Variables: urban, lifeexpf, lifeexpm, literacy, pop_incr, babymort, calories, birth_rt, death_rt, log_gdp, lg_aidsr, b_to_d, fertilty, log_pop, lit_male, lit_fema, log_den

▶ Categorical Variables: region2, religion, climate

▶ Case Labels: country

   Estimation
   ☑ Listwise
   ☑ Pairwise
   ☑ EM
   ☑ Regression

Pairwise means and standard deviations. If a variable has missing data, how much do means and standard deviations of other variables change when the incomplete cases are omitted from the sample? The following panels display estimates of means and standard deviations and also provide additional information about how the pattern of missing values of one variable relates to that of other variables.

Pairwise Means

	URBAN	LIFEEXPF	LITERACY	POP_INCR	BABYMORT	CALORIES	BIRTH_RT	DEATH_RT	LOG_GDP	LG_AIDSR	B_TO_D	FERTILTY	LOG_POP	LIT_FEMA	LOG_DEN
URBAN	**56.5**	70.1	78.3	1.70	42.6	2742	26.0	9.5	3.418	1.39	3.22	3.6	4.115	67.3	1.781
LIFEEXPF	56.5	**70.2**	78.3	1.68	42.3	2754	25.9	9.6	3.422	1.38	3.20	3.6	4.114	67.3	1.784
LIFEEXPM	56.5	70.2	78.3	1.68	42.3	2754	25.9	9.6	3.422	1.38	3.20	3.6	4.114	67.3	1.784
LITERACY	57.0	70.1	**78.3**	1.68	42.7	2742	25.9	9.6	3.413	1.39	3.18	3.6	4.123	67.3	1.789
POP_INCR	56.5	70.2	78.3	**1.68**	42.3	2754	25.9	9.6	3.422	1.38	3.20	3.6	4.114	67.3	1.784
BABYMORT	56.5	70.2	78.3	1.68	**42.3**	2754	25.9	9.6	3.422	1.38	3.20	3.6	4.114	67.3	1.784
CALORIES	54.7	68.8	75.5	1.82	47.0	**2754**	27.5	9.9	3.350	1.55	3.26	3.8	4.219	62.1	1.698
BIRTH_RT	56.5	70.2	78.3	1.68	42.3	2754	**25.9**	9.6	3.422	1.38	3.20	3.6	4.114	67.3	1.784
DEATH_RT	56.4	70.1	78.2	1.69	42.7	2754	26.0	**9.6**	3.418	1.38	3.20	3.6	4.112	67.3	1.775
LOG_GDP	56.5	70.2	78.3	1.68	42.3	2754	25.9	9.6	**3.422**	1.38	3.20	3.6	4.114	67.3	1.784
LG_AIDSR	56.6	70.0	78.0	1.70	43.0	2754	26.2	9.6	3.417	**1.38**	3.21	3.6	4.119	66.9	1.772
B_TO_D	56.4	70.1	78.2	1.69	42.7	2754	26.0	9.6	3.418	1.38	**3.20**	3.6	4.112	67.3	1.775
FERTILTY	56.6	70.0	78.1	1.70	42.9	2754	26.1	9.6	3.417	1.38	3.21	**3.6**	4.116	67.3	1.773
LOG_POP	56.5	70.2	78.3	1.68	42.3	2754	25.9	9.6	3.422	1.38	3.20	3.6	**4.114**	67.3	1.784
LIT_MALE	53.2	67.8	73.6	1.99	51.3	2589	29.0	9.6	3.245	1.34	3.58	3.9	4.142	67.3	1.779
LIT_FEMA	53.2	67.8	73.6	1.99	51.3	2589	29.0	9.6	3.245	1.34	3.58	3.9	4.142	**67.3**	1.779
LOG_DEN	56.5	70.2	78.3	1.68	42.3	2754	25.9	9.6	3.422	1.38	3.20	3.6	4.114	67.3	**1.784**
REGION2	56.2	70.0	78.0	1.70	43.0	2731	26.1	9.6	3.405	1.36	3.23	3.6	4.099	66.9	1.800
RELIGION	56.6	70.2	78.4	1.67	42.3	2754	25.8	9.6	3.421	1.38	3.19	3.6	4.109	67.3	1.786
CLIMATE	56.3	70.0	78.0	1.69	42.8	2754	26.0	9.6	3.415	1.39	3.20	3.6	4.117	66.9	1.772

In the Pairwise Means display, the mean of each column variable is reported for the cases in which the row variable is present (*lifeexpm* and *lit_male* are not shown). The means in which all available values are used are highlighted if you choose *Options* on the Edit menu, then choose *Scripts*, and select *MVA_Table_MOUT_MEAN_Create* before you run the analysis. While means are computed only for quantitative variables, the row variables can be quantitative or categorical. Canada and the United States are the only countries for which *region2* has missing codes. Thus, in the *REGION2* row of the table, when these two countries are omitted from the sample, the mean of *urban* drops from 56.5% to 56.2%, average female life expectancy (*lifeexpf*) drops from 70.2 years to 70 years, the average literacy rate drops from 78.3% to 78%, etc. The average of all infant mortality values (*babymort*) is 42.3. Using only the *babymort* values where *calories* is also present, the mean increases to 47. For the subsample of countries that report both female literacy and infant mortality, the average infant mortality is even greater, 51.3.

Pairwise Standard Deviations

	URBAN	LIFEEXPF	LITERACY	POP_INCR	BABYMORT	CALORIES	BIRTH_RT	DEATH_RT	LOG_GDP	LG_AIDSR	B_TO_D	FERTILTY	LOG_POP	LIT_FEMA	LOG_DEN
URBAN	**24.2**	10.6	22.9	1.19	38.1	562	12.4	4.27	.621	.711	2.13	1.90	.657	28.6	.626
LIFEEXPF	24.2	**10.6**	22.9	1.20	38.1	568	12.4	4.25	.620	.709	2.12	1.90	.654	28.6	.624
LIFEEXPM	24.2	10.6	22.9	1.20	38.1	568	12.4	4.25	.620	.709	2.12	1.90	.654	28.6	.624
LITERACY	23.9	10.7	**22.9**	1.19	38.3	562	12.3	4.27	.623	.714	2.08	1.89	.655	28.6	.623
POP_INCR	24.2	10.6	22.9	**1.20**	38.1	568	12.4	4.25	.620	.709	2.12	1.90	.654	28.6	.624
BABYMORT	24.2	10.6	22.9	1.20	**38.1**	568	12.4	4.25	.620	.709	2.12	1.90	.654	28.6	.624
CALORIES	25.1	11.4	23.1	1.14	38.7	**568**	12.5	4.45	.664	.693	2.10	1.94	.623	27.0	.610
BIRTH_RT	24.2	10.6	22.9	1.20	38.1	568	**12.4**	4.25	.620	.709	2.12	1.90	.654	28.6	.624
DEATH_RT	24.3	10.6	23.0	1.20	38.1	568	12.4	**4.25**	.622	.709	2.12	1.90	.657	28.6	.619
LOG_GDP	24.2	10.6	22.9	1.20	38.1	568	12.4	4.25	**.620**	.709	2.12	1.90	.654	28.6	.624
LG_AIDSR	24.4	10.7	23.1	1.21	38.3	568	12.4	4.27	.627	**.709**	2.14	1.91	.661	28.6	.625
B_TO_D	24.3	10.6	23.0	1.20	38.1	568	12.4	4.25	.622	.709	**2.12**	1.90	.657	28.6	.619
FERTILTY	24.3	10.6	23.1	1.20	38.2	568	12.4	4.26	.624	.709	2.13	**1.90**	.659	28.6	.622
LOG_POP	24.2	10.6	22.9	1.20	38.1	568	12.4	4.25	.620	.709	2.12	1.90	**.654**	28.6	.624
LIT_MALE	24.2	10.7	23.3	1.11	38.3	516	11.9	4.70	.571	.765	2.12	1.88	.671	28.6	.619
LIT_FEMA	24.2	10.7	23.3	1.11	38.3	516	11.9	4.70	.571	.765	2.12	1.88	.671	**28.6**	.619
LOG_DEN	24.2	10.6	22.9	1.20	38.1	568	12.4	4.25	.620	.709	2.12	1.90	.654	28.6	**.624**
REGION2	24.3	10.6	23.0	1.20	38.1	558	12.4	4.29	.613	.700	2.13	1.91	.647	28.6	.615
RELIGION	24.3	10.6	23.0	1.20	38.3	568	12.4	4.27	.623	.713	2.13	1.91	.655	28.6	.626
CLIMATE	24.4	10.6	23.0	1.21	38.2	568	12.4	4.26	.624	.708	2.13	1.91	.658	28.6	.621

The table of Pairwise Standard Deviations has the same structure as that for Pairwise Means. Using all available values, the standard deviation for *urban* is 24.2; restricting the sample to only those countries that also have values of *calories*, the standard deviation increases to 25.1.

Summary panels of mean and standard deviation estimates. Following are the default summary panels of means and standard deviations computed using the listwise, all values, EM, and regression methods. (Results are displayed for requested methods only.)

Summary of Estimates of Means

	URBAN	LIFEEXPF	LITERACY	POP_INCR	BABYMORT	CALORIES	BIRTH_RT	DEATH_RT	LOG_GDP	LG_AIDSR	B_TO_D	FERTILTY	LOG_POP	LIT_FEMA	LOG_DEN
Listwise	49.3	65.6	69.1	2.23	58.6	2570	31.8	9.9	3.107	1.503	3.81	4.3	4.217	61.5	1.665
All Values	56.5	70.2	78.3	1.68	42.3	2754	25.9	9.6	3.422	1.380	3.20	3.6	4.114	67.3	1.784
EM	56.6	70.2	78.4	1.68	42.3	2804	25.9	9.5	3.422	1.377	3.20	3.5	4.114	72.6	1.784
Regression	56.9	70.2	78.5	1.68	42.3	2808	25.9	9.5	3.422	1.369	3.20	3.5	4.114	72.1	1.784

As might be expected, since values are not missing randomly, the listwise estimates stand apart from the others. For urban, the listwise estimate is 49.3%, and for the other methods, the estimates are over 56%; for literacy, the listwise estimate is 69.1%, and for the other methods, the estimates are over 78%; for infant mortality, the listwise estimate is 58.6, and for the other methods, the estimates are under 43, and so on. For most of the variables, the all values, EM, and regression estimates agree fairly well. However, for *calories* and *lit_fema*, the variables with the most values missing, the EM and regression estimates are slightly larger than the all values estimates.

Summary of Estimates of Standard Deviations

	URBAN	LIFEEXPF	LITERACY	POP_INCR	BABYMORT	CALORIES	BIRTH_RT	DEATH_RT	LOG_GDP	LG_AIDSR	B_TO_D	FERTILTY	LOG_POP	LIT_FEMA	LOG_DEN
Listwise	25.2	11.0	22.2	.955	36.7	500.2	11.0	5.0	.542	.745	2.085	1.84	.639	26.85	.617
All Values	24.2	10.6	22.9	1.20	38.1	567.8	12.4	4.3	.620	.709	2.125	1.90	.654	28.61	.624
EM	24.2	10.6	22.8	1.20	38.1	538.3	12.4	4.2	.620	.707	2.117	1.90	.654	28.03	.624
Regression	24.3	10.6	22.7	1.20	38.1	544.0	12.4	4.2	.620	.709	2.116	1.90	.654	27.76	.624

As is true for the previous panel of mean estimates, the listwise estimates shown here differ considerably from the others. The all values, EM, and regression estimates for calories and female literacy fluctuate a little, but there is no definite pattern (that is, neither the EM nor regression estimates appear to reduce the spread more than the other). A regression estimate without random augmentation would undesirably reduce the variance.

Estimates of correlations. By default, when you request the listwise, pairwise, EM, and/or regression methods, the Missing Value procedure, for each method, prints three pivot tables: a panel of means, the covariance matrix, and the correlation matrix. In this section, we omit the means and covariances and display the correlation matrices for the four methods requested.

Missing Data 47

Listwise Correlations

	URBAN	LIFEEXPF	LIFEEXPM	LITERACY	POP_INCR	BABYMORT	CALORIES	BIRTH_RT	DEATH_RT	LOG_GDP	LG_AIDSR	B_TO_D	FERTILTY	LOG_POP	LIT_MALE	LIT_FEMA	LOG_DEN
URBAN	1.00																
LIFEEXPF	.74	1.00															
LIFEEXPM	.71	.99	1.00														
LITERACY	.61	.82	.78	1.00													
POP_INCR	-.17	-.38	-.31	-.55	1.00												
BABYMORT	-.70	-.95	-.93	-.89	.40	1.00											
CALORIES	.67	.71	.70	.56	-.37	-.69	1.00										
BIRTH_RT	-.56	-.81	-.77	-.82	.77	.80	-.64	1.00									
DEATH_RT	-.59	-.85	-.87	-.62	.01	.77	-.45	.55	1.00								
LOG_GDP	.79	.76	.73	.63	-.33	-.74	.79	-.66	-.52	1.00							
LG_AIDSR	-.13	-.30	-.37	-.01	-.13	.15	-.24	.14	.50	-.04	1.00						
B_TO_D	.28	.30	.35	.07	.68	.08	.17	-.58	.13	-.31	1.00						
FERTILTY	-.52	-.78	-.74	-.81	.75	.78	-.57	.97	.57	-.57	.16	.13	1.00				
LOG_POP	-.18	.00	.02	.04	-.27	.03	.08	-.18	-.07	-.23	-.34	-.25	-.21	1.00			
LIT_MALE	.59	.75	.72	.94	-.53	-.80	.56	-.75	-.55	.57	-.09	.05	-.75	.15	1.00		
LIT_FEMA	.63	.81	.77	.96	-.57	-.85	.53	-.80	-.59	.59	-.03	.06	-.81	.07	.96	1.00	
LOG_DEN	-.22	.01	.03	.01	-.22	-.05	.00	-.15	.01	-.21	-.18	-.08	-.15	.30	.11	.08	1.00

Pairwise Correlations

	URBAN	LIFEEXPF	LIFEEXPM	LITERACY	POP_INCR	BABYMORT	CALORIES	BIRTH_RT	DEATH_RT	LOG_GDP	LG_AIDSR	B_TO_D	FERTILTY	LOG_POP	LIT_MALE	LIT_FEMA	LOG_DEN
URBAN	1.00																
LIFEEXPF	.74	1.00															
LIFEEXPM	.73	.98	1.00														
LITERACY	.65	.87	.81	1.00													
POP_INCR	-.37	-.58	-.50	-.70	1.00												
BABYMORT	-.72	-.96	-.94	-.90	.60	1.00											
CALORIES	.69	.78	.77	.68	-.61	-.78	1.00										
BIRTH_RT	-.63	-.86	-.80	-.87	.86	.87	-.76	1.00									
DEATH_RT	-.48	-.70	-.74	-.49	-.04	.63	-.35	.37	1.00								
LOG_GDP	.75	.83	.80	.73	-.56	-.82	.85	-.77	-.40	1.00							
LG_AIDSR	-.06	-.15	-.17	-.01	-.07	.04	-.07	.05	.34	.06	1.00						
B_TO_D	-.03	-.09	-.01	-.27	.80	.12	-.24	.48	-.51	-.21	-.22	1.00					
FERTILTY	-.62	-.84	-.78	-.87	.84	.83	-.70	.98	.40	-.69	.07	.45	1.00				
LOG_POP	-.14	-.09	-.08	-.05	-.08	.11	.05	-.03	.01	-.22	-.21	-.15	-.06	1.00			
LIT_MALE	.59	.78	.72	.95	-.62	-.81	.58	-.79	-.49	.61	-.14	-.15	-.80	.08	1.00		
LIT_FEMA	.61	.82	.74	.97	-.64	-.84	.55	-.83	-.51	.63	-.10	-.15	-.84	.00	.96	1.00	
LOG_DEN	.02	.13	.15	.08	-.25	-.15	.05	-.22	-.06	.00	-.14	-.11	-.22	.14	.14	.11	1.00

EM Correlations[a]

	URBAN	LIFEEXPF	LIFEEXPM	LITERACY	POP_INCR	BABYMORT	CALORIES	BIRTH_RT	DEATH_RT	LOG_GDP	LG_AIDSR	B_TO_D	FERTILTY	LOG_POP	LIT_MALE	LIT_FEMA	LOG_DEN
URBAN	1.00																
LIFEEXPF	.74	1.00															
LIFEEXPM	.73	.98	1.00														
LITERACY	.65	.86	.81	1.00													
POP_INCR	-.38	-.58	-.50	-.70	1.00												
BABYMORT	-.72	-.96	-.94	-.90	.60	1.00											
CALORIES	.66	.75	.74	.64	-.58	-.76	1.00										
BIRTH_RT	-.63	-.86	-.80	-.87	.86	.87	-.74	1.00									
DEATH_RT	-.48	-.70	-.74	-.47	-.04	.63	-.32	.37	1.00								
LOG_GDP	.75	.83	.80	.73	-.56	-.82	.81	-.77	-.40	1.00							
LG_AIDSR	-.06	-.15	-.17	-.02	-.06	.05	-.08	.05	.34	.06	1.00						
B_TO_D	-.04	-.09	-.01	-.28	.80	.12	-.26	.48	-.51	-.21	-.22	1.00					
FERTILTY	-.61	-.84	-.79	-.86	.84	.83	-.67	.98	.40	-.69	.07	.45	1.00				
LOG_POP	-.14	-.09	-.08	-.04	-.08	.11	.00	-.03	.01	-.22	-.20	-.15	-.06	1.00			
LIT_MALE	.63	.80	.75	.95	-.68	-.83	.64	-.82	-.41	.67	-.06	-.31	-.82	.07	1.00		
LIT_FEMA	.65	.84	.78	.97	-.70	-.86	.60	-.86	-.43	.68	-.03	-.31	-.86	.01	.96	1.00	
LOG_DEN	.02	.13	.15	.09	-.25	-.15	.10	-.22	-.07	.00	-.14	-.11	-.23	.14	.16	.14	1.00

[a]. Little's MCAR test: Chisquare=157.252, df=108, Prob=.001

Little's chi-square test for missing completely at random is printed with EM results. Here, the chi-square is 157.3 ($df = 108$ and p value = 0.001), agreeing with the nonrandom pattern of missing values identified in the displays above.

When the regression method of estimation is requested and the number of complete cases is less than half the total number of cases, the Missing Value procedure augments each estimated value with a normal deviate instead of the residual from a randomly selected complete case. For these data, 53% of the cases are complete, and we would be more comfortable if more cases were complete. In the next section, we will examine the element-by-element differences between correlation matrices estimated with regression methods using random normal deviates and random residuals.

Regression Correlations[a]

	URBAN	LIFEEXPF	LIFEEXPM	LITERACY	POP_INCR	BABYMORT	CALORIES	BIRTH_RT	DEATH_RT	LOG_GDP	LG_AIDSR	B_TO_D	FERTILTY	LOG_POP	LIT_MALE	LIT_FEMA	LOG_DEN
URBAN	1.00																
LIFEEXPF	.74	1.00															
LIFEEXPM	.73	.98	1.00														
LITERACY	.64	.87	.81	1.00													
POP_INCR	-.38	-.58	-.50	-.69	1.00												
BABYMORT	-.72	-.96	-.94	-.90	.60	1.00											
CALORIES	.65	.76	.76	.62	-.54	-.76	1.00										
BIRTH_RT	-.63	-.86	-.80	-.86	.86	.87	-.73	1.00									
DEATH_RT	-.48	-.70	-.74	-.48	-.04	.63	-.37	.37	1.00								
LOG_GDP	.75	.83	.80	.73	-.56	-.82	.82	-.77	-.40	1.00							
LG_AIDSR	-.08	-.15	-.18	-.03	-.06	.05	-.07	.06	.34	.05	1.00						
B_TO_D	-.05	-.09	-.01	-.27	.80	.12	-.21	.48	-.51	-.21	-.22	1.00					
FERTILTY	-.61	-.84	-.79	-.85	.84	.83	-.67	.98	-.03	-.69	.08	.46	1.00				
LOG_POP	-.14	-.09	-.08	-.05	-.08	.11	.00	-.03	.01	-.22	-.21	-.15	-.05	1.00			
LIT_MALE	.57	.78	.73	.92	-.64	-.80	.56	-.78	-.43	.63	-.08	-.26	-.78	.05	1.00		
LIT_FEMA	.63	.83	.77	.96	-.69	-.84	.57	-.84	-.44	.67	-.03	-.29	-.85	.00	.96	1.00	
LOG_DEN	.02	.13	.15	.09	-.25	-.15	.07	-.22	-.07	.00	-.15	-.11	-.23	.14	.13	.12	1.00

a. Residual of randomly picked case added to each estimate.

Comparing Estimates of Covariance and Correlation Matrices

In a large study, it is difficult to compare two correlation matrices for differences (or to determine whether they differ at all). We saved four correlation matrices shown above and also a matrix of regression estimates augmented with random normal deviates. We used SPSS's MATRIX procedure to compute the difference between elements in each pair of matrices.

You may not need to do this, but following are the steps we took to accomplish the task:
- In the Output Navigator, click on the first matrix to select it.
- To remove row and column labels choose:

  Utilities
    Run Script...
    ☑ Remove labels.sbs (or remlabel.sbs)

  Click Run.
- Open the Syntax Editor:

  File
    New
      Syntax
- Return to the Output Navigator and click and hold the matrix, and drag it to the open Syntax Editor.
- Type the following syntax before the matrix:

  ```
  MATRIX DATA VARIABLES = urban, lifeexpf, lifeexpm, literacy, pop_incr,
                          babymort, calories, birth_rt, death_rt,
                          log_gdp, lg_aidsr, b_to_d, fertility,
                          log_pop, lit_male, lit_fema, log_den.
  BEGIN DATA.
  ```
- Type the following syntax after the matrix:

  ```
  END DATA.
  SAVE OUTFILE=corr_list.sav.
  ```
- Repeat the above steps for one or more other matrices, entering a SAVE command after each. For example,

  ```
  SAVE OUTFILE=corr_em.sav.
  SAVE OUTFILE=corr_resid.sav.
  SAVE OUTFILE=corr_norm.sav.
  ```
- After the last SAVE command, add syntax for the MATRIX procedure:

  ```
  MATRIX.
  GET list  / FILE=corr_list.sav.
  GET em    / FILE=corr_em.sav.
  GET resid / FILE=corr_resid.sav.
  GET norm  / FILE=corr_norm.sav.

  PRINT (list - em)    / FORMAT='F4.2' / TITLE="List - EM".
  PRINT (resid - norm) / FORMAT='F4.2' / TITLE="Resid - Norm".

  END MATRIX.
  ```

The matrix of element-by-element differences between the correlation matrices estimated by the listwise and EM methods is displayed in Figure 2.8. (Two columns of zeros are deleted.) The order of variables is the same as that in the preceding correlation matrices. For example, differences for variables correlated with *calories* are in the seventh row and the seventh column. Differences between correlations involving *log_den* are in the last row, and those correlated with male and female literacy are in the two rows preceding *log_den*. *Babymort* is in the sixth row and column.

Figure 2.8   Differences between listwise and EM estimates of correlations

```
 .00
-.01  .00                       babymort
-.02  .01  .00                     |
-.04 -.04 -.03  .00                |
 .20  .20  .20  .14  .00           |
 .02  .01  .01  .01 -.20  .00
 .01 -.05 -.04 -.08  .22  .07  .00
 .07  .05  .04  .05 -.09 -.06  .10  .00
-.11 -.16 -.13 -.15  .05  .14 -.13  .18  .00
 .03 -.07 -.07 -.10  .22  .09 -.03  .11 -.11  .00
-.06 -.15 -.20  .01 -.07  .10 -.16  .09  .16 -.10  .00
 .32  .39  .36  .35 -.11 -.38  .34 -.32 -.07  .34 -.10  .00  ──── b_to_d
 .08  .05  .04  .05 -.09 -.06  .11 -.01  .17  .12  .09 -.32  .00
-.04  .09  .10  .09 -.19 -.07  .08 -.16 -.09 -.01 -.14 -.09 -.15  .00
-.04 -.05 -.03 -.01  .15  .02 -.07  .07 -.15 -.10 -.02  .36  .07  .08  .00
-.03 -.03 -.01 -.01  .14  .00 -.07  .05 -.16 -.10  .00  .36  .05  .07 -.01  .00
-.24 -.11 -.12 -.08  .03  .11 -.10  .07  .08 -.21 -.04  .03  .08  .16 -.06 -.05  .00
```

As might be expected, since values are not missing randomly, the estimates differ markedly, especially for *b_to_d*, the ratio of births to deaths, in the twelfth row and column. For example, the listwise estimate of the correlation between *b_to_d* and *babymort* is −0.26; the EM estimate is 0.12—making a difference of −0.38 (the pairwise and regression estimates are also 0.12). The data for *babymort* are complete; for *b_to_d*, one case is missing. In the earlier search for nonrandom patterns, *b_to_d* was not noticed. The plots in Figure 2.9 provide another view.

Figure 2.9 Scatterplots highlighting the pattern of listwise missing

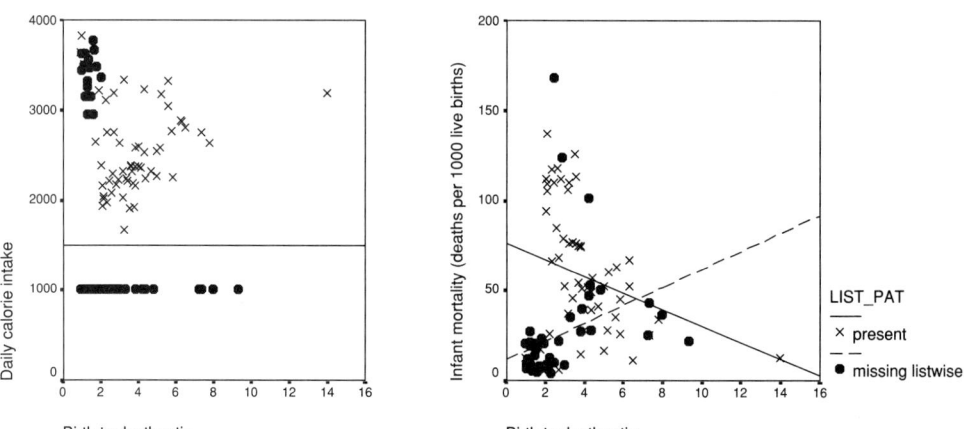

In the plot on the left, we arbitrarily assigned 1000 to the values of *calories* that are missing (they fall below the horizontal line). It is easy to see that the distribution of *b_to_d* values that exists when *calories* is missing differs little from that marked by *X*'s above the line. Now look at the cluster of filled circles at the top of the plot—these are the values of *calories* and *b_to_d* omitted by listwise deletion. They certainly are not a random subsample from the bivariate distribution of *babymort* and *b_to_d*. In the plot on the right, the complete data for the pair (*x*'s) and those omitted by listwise deletion are shown. If we add a line-of-best-fit to each group, the line for the missing listwise cases has a positive slope, while that for the complete pairs has a negative slope.

Most of the differences (not shown) between correlations estimated by the EM and pairwise methods are 0 except for the rows and columns involving correlations with calories and the male and female literacy rates. Where differences exist, they are smaller than those in Figure 2.8. The differences for correlations between *b_to_d* and the male and female literacy rates (0.15) are two times as large as any other difference in the matrix. All other differences involving *b_to_d* are 0 or 0.01.

The only differences (not shown) between correlations estimated by 1) EM and regression with random residuals and 2) EM and regression with random normal variates involve calories and the male and female literacy rates. Thus, it is not surprising to see in Figure 2.10 that the differences between correlation estimates computed via the two regression methods are small. We are unable to say which estimates are *best* because we do not know the underlying truth. The values are gone. The only thing we can conclude is that the listwise estimates are the worst because the data clearly are not missing randomly. In the next example, we will examine values imputed via the different methods.

Figure 2.10 Differences between regression estimates augmented with random residuals and those with random normal deviates

```
.00
.00  .00                    babymort
.00  .00  .00
.01  .00  .00  .00
-.01 .00  .00  .00  .00
.00  .00  .00  .00  .00  .00
.01 -.02 -.02 -.02  .02  .02  .00
-.01 .00  .00  .00  .00  .00  .02  .00
.01  .00  .00  .00  .00  .00  .00  .00  .00
.00  .00  .00 -.01  .00  .00 -.02  .00  .00  .00
.00  .02  .02  .02 -.02 -.02 -.06 -.02 -.02  .01  .00
-.01 .00  .00 -.01  .00  .00  .02  .00  .00  .00 -.01  .00 ——————— b_to_d
-.01 .00  .00 -.01  .00  .00  .03  .00  .00  .00 -.03  .00  .00
.00  .00  .00  .01  .00  .00  .00  .00  .00  .00  .00  .00  .00  .00
-.05 .01  .01  .01  .03 -.01  .01  .02 -.03  .04  .05  .03  .02 -.01  .00
.01  .00  .01  .00 -.02 -.01  .00 -.01  .02  .02  .03 -.03 -.02  .00 -.03  .00
.00  .00  .00  .01  .00  .00  .02  .00  .00  .00  .03  .00  .00  .00 -.01  .01  .00
```

Example 4:
Estimating Replacement Values: Imputation

The Missing Value procedure provides EM and regression methods for estimating (imputing) replacement values, but this should not be done until the data have been screened for recording errors and variables in need of a symmetrizing transformation.

To save the filled-in data, select *Save completed data* in the EM or Regression dialog box when you specify the estimation procedure (estimation is described in Example 4). In one run, you can save a file with completed data from an EM method and another file from a regression method but not more than one file from a single method. For the examples in this section, we use data files saved from the default EM and regression methods described in the last example.

Values in the *world95m* data are not randomly missing (we're sure that they are not missing *completely* at random and also have doubts about satisfying the MAR condition). So, how good are the imputed values? In this section, we display some plots that you might create when evaluating your own filled-in data. You can:

- Display the variables with the most values missing in a pair of bivariate scatterplots with the same plot scales—one using the observed data only and the other using the imputed values. For our example, we use *calories* and *lit_fema*.
- For the same variable, plot the imputed values from one method against those from another. For female literacy, we plot imputed values from the regression method with random residuals against those from the EM method.
- Using knowledge of the subject matter, design displays that highlight the presence of the observed and imputed values.

Generating pattern variables. In the plots that follow, pattern variables are used as case selection variables to group and identify observed and imputed values. To generate pattern variables for calories and female literacy, we use *Compute* on the Transform menu with its MISSING function to form two 0,1 variables (the values for each new variable are 0 for missing and 1 for present). The SPSS statements for doing this are *pat_calr* = 1 − MISSING(*calories*) and *pat_litf* = 1 − MISSING(*lit_fema*).

We also generate a third pattern variable that combines the missing/present information for calories and female literacy by specifying *pat_both* = 10*pat_calr* + *pat_litf*. The result of this transformation is four codes: 0, 1, 10, and 11. For example, if, for a case, both values are present (*pat_calr* and *pat_litf* are both 1), the value of the new variable *pat_both* is 10*1 + 1 or 11. When only female literacy is missing, the code for *pat_both* is 10; when only calories is missing, the code is 1; and when values of both variables are missing, the code is 0.

Scatterplots of observed and imputed values. In some plots below, we use *Select Cases* on the Data menu to select countries (cases) in which values of both calories and female literacy are present (*pat_both* = 11), and in other plots, countries in which one variable is missing or both are missing (*pat_both* is less than 11).

Following are some of the chart features we use (click on the graph and choose *SPSS Chart Object* on the Edit menu to access the Chart Editor):

- To set minimum and maximum limits, increments, and grids, use *Axis* on the Chart menu in the Chart Editor.

- To set the position of a reference line, select *Reference Line* on the Chart menu. For literacy, we add a line at 100% to see how many imputed values fall above the valid range.

- To select distinct symbols for cases missing literacy only, calories only, and both variables, click on a plot point in the first group, select the *Marker* button on the Chart Editor toolbar, select the symbol and plot size you want, and, finally, select *Apply*, not *Apply All*. Then, repeat the selections for each group.

The observed values of female literacy and calories are plotted in the left frame in Figure 2.11. They are, of course, the same for both imputation methods.

Figure 2.11 Observed and imputed values of calories and female literacy

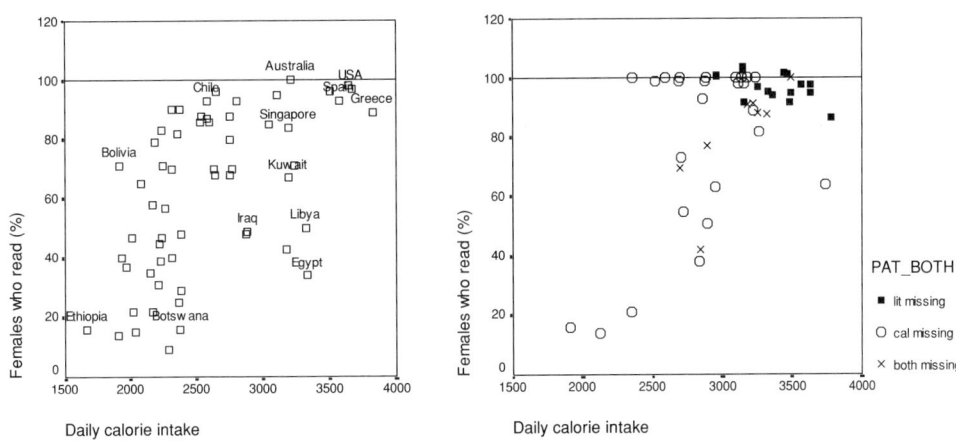

Values imputed via the EM method are displayed in the right frame in Figure 2.11. Notice that the plot scales are the same, and when female literacy is missing (dark squares), its imputed values tend to be high. Some of these imputed values even fall above 100%. In Figure 0.5, by adding country names to identify these countries, we find Japan, UK, and Germany.

Figure 2.12 EM imputed values with country names

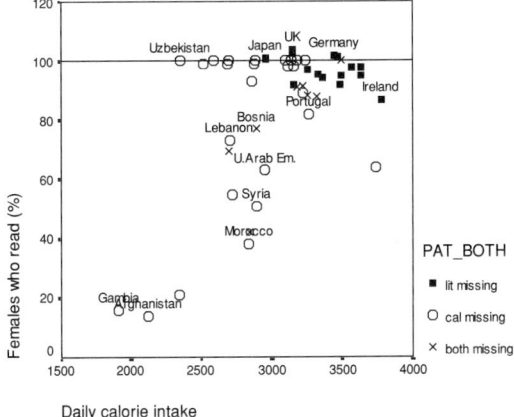

56   Chapter 2, Example 5

Values imputed by the regression method are plotted in Figure 2.13. Iceland's estimated female literacy is considerably above 100%.

Figure 2.13 Values imputed by the regression method

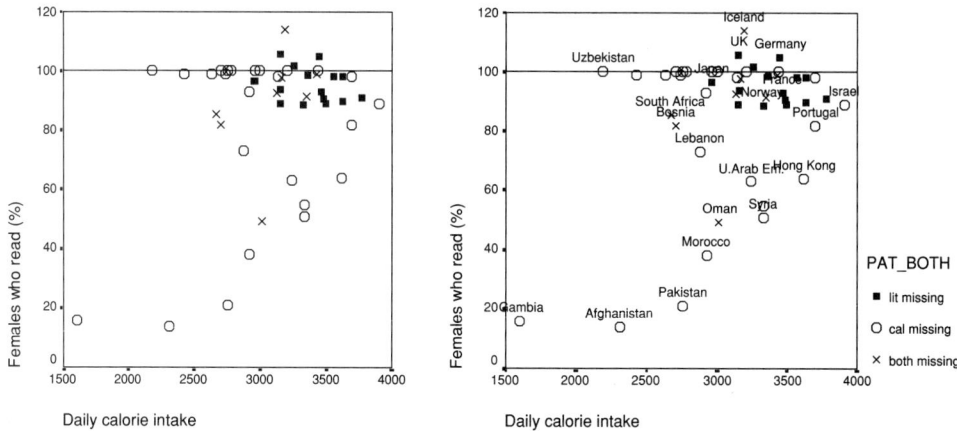

Comparing values imputed by the EM and regression methods. In Figure 2.14, for female literacy, values imputed by the EM method are compared with those imputed by the regression method with random residuals. The Add Variables dialog box under *Merge Files* on the Data menu was used to merge the two files of imputed values side-by-side.

Ideally, the plot points should fall along a line connecting the intersection of grid lines for the same percentage (for example, 80% for EM with 80% for regression). When both calories and female literacy are estimated (the plot symbol *X* marks Oman, Bosnia, South Africa, and Iceland), the regression estimates tend to be higher than the EM estimates. The points with estimated literacy values (small dark squares) are clustered together, making it difficult to identify them.

Figure 2.14 EM and regression imputed values for female literacy

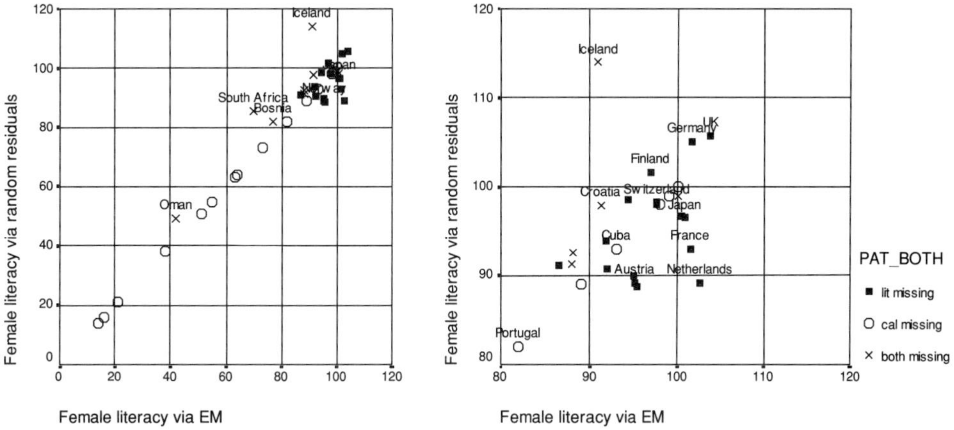

[Two scatter plots comparing Female literacy via EM (x-axis) vs Female literacy via random residuals (y-axis). Left plot shows full range 0–120; right plot zooms into 80–120 range. Legend: PAT_BOTH — ■ lit missing, ○ cal missing, × both missing. Countries labeled include Iceland, Japan, South Africa, Norway, Bosnia, Oman, Germany, UK, Finland, Croatia, Switzerland, Japan, Cuba, France, Austria, Netherlands, Portugal.]

On the right side in Figure 2.14, we zoom in on this area of the plot, finding that the largest discrepancies between the methods are for Iceland and the Netherlands. Iceland's *x-y* plot coordinates are (91%, 114%) and the Netherlands' are (103%, 89%).

In Figure 2.15, we compare imputed values for calories. The regression filled-in value for Israel is almost 700 calories larger than the EM value (3908 calories versus 3223). In general, when there is a difference, the regression estimates tend to be higher more often than they are lower.

Figure 2.15 EM and regression imputed values for calories

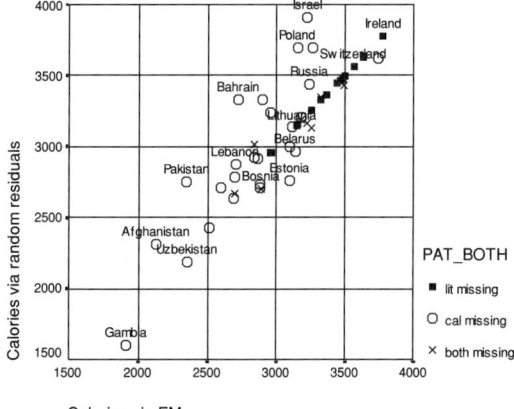

Using the subject matter to design displays. In Example 1 and Example 2, it was noted that the pattern of missing data varied by geographical region. Neither imputation method makes an adjustment for these subpopulation differences. The left frame in Figure 2.16 is a scatterplot of the observed female literacy values (open circles) and the EM imputed values (filled squares) against the code for geographical region (code 1 is Europe, ..., code 6 is Latin America).

Figure 2.16 Observed and imputed values of female literacy and calories grouped by region

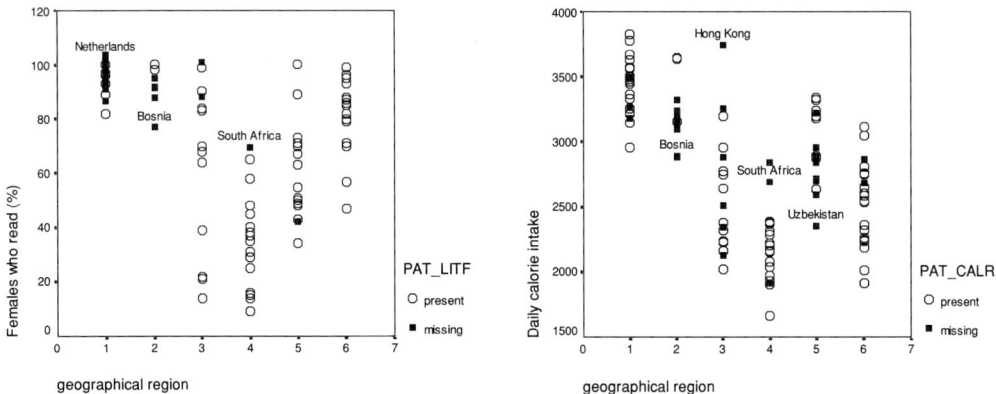

Instead of female literacy, observed and imputed values of calories are displayed in the right frame in Figure 2.16. The estimate for Hong Kong is much higher than the observed values in the Pacific/Asia group (code 3). Visually, Hong Kong looks as though it might be a member of the European region (code 1). This is not as unreasonable as it might seem at first, because Hong Kong's infant mortality rate, female life expectancy, GDP per capita, and high proportion of people living in cities *are* like those of European countries.

Figure 2.17 Female literacy versus literacy

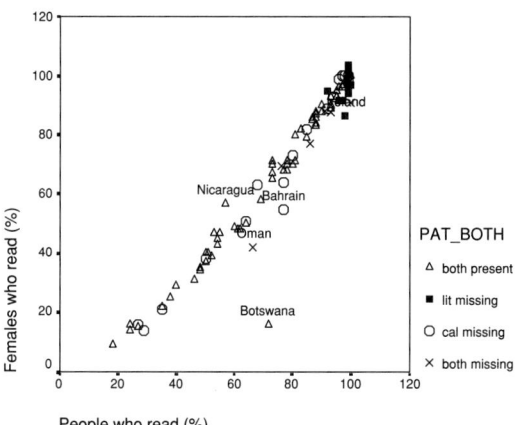

Another internal cross-check is to compare observed and imputed values of female literacy with the overall literacy rate (it has only two values missing). Look at Botswana in Figure 2.17, where neither value is imputed. The original data screening was not thorough enough. Several sources report that the literacy rate for this older and more prosperous country among countries in Africa is 72% or 74%. Where did the value of 16% for female literacy come from? Is it a recording error? Did someone use total population size when computing the rate instead of the number of females? Also, does the presence of this outlier distort the other estimates?

Multiple Imputation

Multiple imputation is a technique that replaces each missing or deficient value with two or more values simulated from a suitable distribution. Options with the regression method that incorporate randomness allow you to perform a variation of multiple imputation. In the Missing Value procedure, imputed data values can be augmented with a random normal (0, RMS) deviate, a random t scaled by the square root of RMS (with 5 or a user-specified degrees of freedom), or a residual randomly selected from another case.

Multiple imputation is accomplished by generating m, say 7, data files via the regression method (the seed for the random number generator changes for each file) and performing the desired statistical analysis (for example, regression) with each completed data set. The final estimate of each parameter is the average of the respective estimates from the individual runs, and the multiple imputation estimate of the covariance matrix is the sum of the pooled *within* components plus a *between* component obtained from deviations between the final parameter estimate and the individual estimates.

Syntax Reference

MVA

```
MVA [VARIABLES=] {varlist}
                {ALL    }

  [/CATEGORICAL=varlist]

  [/MAXCAT={25**}]
          {n   }

  [/ID=varname]
```

Description:

```
    [/NOUNIVARIATE]

    [/TTEST [PERCENT={5}] [{T  }] [{DF  }] [{PROB  }] [{COUNTS  }] [{MEANS  }]]
                    {n}   {NOT}   {NODF}   {NOPROB}]  {NOCOUNTS}   {NOMEANS}

    [/CROSSTAB [PERCENT={5}]]
                       {n}

    [/MISMATCH [PERCENT={5}] [NOSORT]]
                        {n}

    [/DPATTERN [SORT=varname[({ASCENDING })] [varname ... ]]
                             {DESCENDING}
               [DESCRIBE=varlist]]

    [/MPATTERN [NOSORT] [DESCRIBE=varlist]]

    [/TPATTERN [NOSORT] [DESCRIBE=varlist] [PERCENT={1}]]
                                                    {n}
```

Estimation:

```
    [/LISTWISE]

    [/PAIRWISE]

    [/EM  [predicted_varlist] [WITH predictor_varlist]
        [([TOLERANCE={0.001}  ]
                     {value}
          [CONVERGENCE={0.0001}]
                       {value }
          [ITERATIONS={25}     ]
                      {n }
          [TDF=n             ]
          [LAMBDA=a          ]
          [PROPORTION=b      ]
          [OUTFILE='file'    ])]
```

```
[/REGRESSION  [predicted_varlist] [WITH predictor_varlist]
             [([TOLERANCE={0.001}                             ]
                          {n    }
              [FLIMIT={4.0}                                   ]
                      {N  }
              [NPREDICTORS=number_of_predictor_variables]
              [ADDTYPE={RESIDUAL*}                            ]
                       {NORMAL   }
                       {T[({5})  }
                          {n}
                       {NONE     }
              [OUTFILE='file'                          ])]]
```

* If the number of complete cases is less than half the number of cases, the default **ADDTYPE** specification is **NORMAL**.

** Default if the subcommand is omitted.

Examples:
```
MVA VARIABLES=populatn density urban religion lifeexpf region
  /CATEGORICAL=region
  /ID=country
  /MPATTERN DESCRIBE=region religion.

MVA VARIABLES=all
  /EM males msport WITH males msport gradrate facratio.
```

Overview

MVA (Missing Value Analysis) describes the missing value patterns in a data file (data matrix). It can estimate the means, the covariance matrix, and the correlation matrix by using listwise, pairwise, regression, and EM estimation methods. Missing values themselves can be estimated (imputed), and you can then save the new data file.

Options

Categorical variables. String variables are automatically defined as categorical. For a long string variable, only the first eight characters are used to define categories. Quantitative variables can be designated as categorical by using the CATEGORICAL subcommand.

MAXCAT specifies the maximum number of categories for any categorical variable. If any categorical variable has more than the specified number of distinct values, MVA is not executed.

Analyzing patterns. For each quantitative variable, the TTEST subcommand produces a series of t tests. Values of the quantitative variable are divided into two groups, based on the presence or absence of other variables. These pairs of groups are compared using the t test.

Crosstabulating categorical variables. The CROSSTAB subcommand produces a table for each categorical variable, showing, for each category, how many nonmissing values are in the other variables and the percentages of each type of missing value.

Displaying patterns. DPATTERN displays a case-by-case data pattern with codes for system-missing, user-missing, and extreme values. MPATTERN displays only the cases that have missing values and sorts by the pattern formed by missing values. TPATTERN tabulates the cases that have a common pattern of missing values. The pattern tables have sorting options. Also, descriptive variables can be specified.

Labeling cases. For pattern tables, an ID variable can be specified to label cases.

Suppression of rows. To shorten tables, the PERCENT keyword suppresses missing value patterns that occur relatively infrequently.

Statistics. Displays of univariate, listwise, and pairwise statistics are available.

Estimation. EM and REGRESSION use different algorithms to supply estimates of missing values, which are used in calculating estimates of the mean vector, the covariance matrix, and the correlation matrix of dependent variables. The estimates can be saved as replacements for missing values in a new data file.

Basic Specification

The basic specification depends on whether you want to describe the missing data pattern or estimate statistics. Often, description is done first, and then, considering the results, an estimation is done. Alternatively, both description and estimation can be done by using the same MVA command.

Descriptive analysis. A basic descriptive specification includes a list of variables and a statistics or pattern subcommand. For example, a list of variables and the subcommand DPATTERN would show missing value patterns for all cases with respect to the list of variables.

Estimation. A basic estimation specification includes a variable list and an estimation method. For example, if the EM method is specified, SPSS estimates the mean vector, the covariance matrix, and the correlation matrix of quantitative variables with missing values.

Syntax Rules

- A variables specification is required directly after the command name. The specification can be either a variable list or the keyword ALL.
- The CATEGORICAL, MAXCAT, and ID subcommands, if used, must be placed after the variables list and before any other subcommand. These three subcommands can be in any order.
- Any combination of description and estimation subcommands can be specified. For example, both the EM and REGRESSION subcommands can be specified in one MVA command.
- Univariate statistics are displayed unless the NOUNIVARIATE subcommand is specified. Thus, if only a list of variables is specified, with no description or estimation subcommands, univariate statistics are displayed.
- If a subcommand is specified more than once, only the last one is honored.
- The following words are reserved as keywords or internal commands in the MVA procedure: VARIABLES, SORT, NOSORT, DESCRIBE, and WITH. They cannot be used as variable names in MVA.
- The tables *Summary of Estimated Means* and *Summary of Estimated Standard Deviations* are produced if you specify more than one way to estimate means and standard deviations. The methods include univariate (default), listwise, pairwise, EM, and regression. For example, these tables are produced when you specify both LISTWISE and EM.

Symbols

The symbols displayed in the DPATTERN and MPATTERN table cells are:

+   Extremely high value

-     Extremely low value

S   System-missing value

A   First type of user-missing value

B   Second type of user-missing value

C   Third type of user-missing value

- An extremely high value is more than 1.5 times the interquartile range above the 75th percentile, if (number of variables) $\times n \log n \leq 150000$, where n is the number of cases.
- An extremely low value is more than 1.5 times the interquartile range below the 25th percentile, if (number of variables) $\times n \log n \leq 150000$, where n is the number of cases.
- For larger files—that is, (number of variables) $\times n \log n > 150000$—extreme values are two standard deviations from the mean.

Missing Indicator Variables

For each variable in the VARIABLES list, a binary indicator variable is formed (internal to MVA), indicating whether a value is present or missing.

VARIABLES Subcommand

A list of variables or the keyword ALL is required.

- The order in which the variables are listed determines the default order in the output.
- The keyword VARIABLES is optional.
- If the keyword ALL is used, the default order is the order of variables in the working data file.
- String variables specified in the variable list, whether short or long, are automatically defined as categorical. For a long string variable, only the first eight characters of the values are used to distinguish categories.
- The list of variables must precede all other subcommands.
- Multiple lists of variables are not allowed.

CATEGORICAL Subcommand

The MVA procedure automatically treats all string variables in the variables list as categorical. You can designate numeric variables as categorical by listing them on the CATEGORICAL subcommand. If a variable is designated categorical, it will be ignored if listed as a dependent or independent variable on the REGRESSION or EM subcommand.

MAXCAT Subcommand

The MAXCAT subcommand sets the upper limit of the number of distinct values that each categorical variable in the analysis can have. The default is 25. This limit affects string variables in the variables list and also the categorical variables defined by the CATEGORICAL subcommand. A large number of categories can slow the analysis considerably. If any categorical variable violates this limit, MVA does not run.

Example
```
MVA VARIABLES=populatn density urban religion lifeexpf region
  /CATEGORICAL=region
  /MAXCAT=30
  /MPATTERN.
```

- The CATEGORICAL subcommand specifies that *region*, a numeric variable, is categorical. The variable *religion*, a string variable, is automatically categorical.
- The maximum number of categories in *region* or *religion* is 30. If either has more than 30 distinct values, MVA produces only a warning.
- Missing data patterns are shown for those cases that have at least one missing value in the specified variables.
- The summary table lists the number of missing and extreme values for each variable, including those with no missing values.

ID Subcommand

The ID subcommand specifies a variable to label cases. These labels appear in the patterns tables. Without this subcommand, the SPSS case numbers are used.

Example
```
MVA VARIABLES=populatn density urban religion lifeexpf region
  /CATEGORICAL=region
  /MAXCAT=20
  /ID=country
  /MPATTERN.
```

- The values of the variable *country* are used as case labels.
- Missing data patterns are shown for those cases that have at least one missing value in the specified variables.

NOUNIVARIATE Subcommand

By default, MVA computes univariate statistics for each variable—the number of cases with nonmissing values, the mean, the standard deviation, the number and percentage of missing values, and the counts of extreme low and high values. (Means, standard deviations, and extreme value counts are not reported for categorical variables.)

- To suppress the univariate statistics, specify NOUNIVARIATE.

Examples

```
MVA VARIABLES=populatn density urban religion lifeexpf region
    /CATEGORICAL=region
    /CROSSTAB PERCENT=0.
```

- Univariate statistics (number of cases, means, and standard deviations) are displayed for *populatn*, *density*, *urban*, and *lifeexpf*. Also, the number of cases, counts and percentages of missing values, and counts of extreme high and low values are displayed.
- The total number of cases and counts and percentages of missing values are displayed for *region* and *religion* (a string variable).
- Separate crosstabulations are displayed for *region* and *religion*.

```
MVA VARIABLES=populatn density urban religion lifeexpf region
    /CATEGORICAL=region.
    /NOUNIVARIATE
    /CROSSTAB PERCENT=0.
```

- Only crosstabulations are displayed, no univariate statistics.

TTEST Subcommand

For each quantitative variable, a series of *t* tests is computed to test the difference of means between two groups defined by a missing indicator variable for each of the other variables (see "Missing Indicator Variables" on p. 66). For example, a *t* test is performed on *populatn* between two groups defined by whether their values are present or missing for *calories*. Another *t* test is performed on *populatn* for the two groups defined by whether their values for *density* are present or missing, and so on for the remainder of the variable list.

PERCENT=n   *Omit indicator variables with less than the specified percentage of missing values.* You can specify a percentage from 0 to 100. The default is 5, indicating the omission of any variable with less than 5% missing values. If you specify 0, all rows are displayed.

Display of Statistics

The following statistics can be displayed for a *t* test:

- The **t statistic**, for comparing the means of two groups defined by whether the indicator variable is coded as missing or nonmissing (see "Missing Indicator Variables" on p. 66).

    T   *Display the t statistics.* This is the default.

    NOT   *Suppress the t statistics.*

- The **degrees of freedom** associated with the *t* statistic.

    DF   *Display the degrees of freedom.* This is the default.

    NODF   *Suppress the degrees of freedom.*

- The **probability** (two-tailed) associated with the *t* test, calculated for the variable tested without reference to other variables. Care should be taken when interpreting this probability.

    PROB   *Display probabilities.*

    NOPROB   *Suppress probabilities.* This is the default.

- The **number of values in each group**, where groups are defined by values coded as missing and present in the indicator variable.

    COUNTS   *Display counts.* This is the default.

    NOCOUNTS *Suppress counts.*

- The **means** of the groups, where groups are defined by values coded as missing and present in the indicator variable.

    MEANS   *Display means.* This is the default.

    NOMEANS   *Suppress means.*

Example

```
MVA VARIABLES=populatn density urban religion lifeexpf region
  /CATEGORICAL=region
  /ID=country
  /TTEST.
```

- The TTEST subcommand produces a table of *t* tests. For each quantitative variable named in the variables list, a *t* test is performed, comparing the mean of the values for which the other variable is present against the mean of the values for which the other variable is missing.
- The table displays default statistics, including values of *t*, degrees of freedom, counts, and means.

CROSSTAB Subcommand

CROSSTAB produces a table for each categorical variable, showing the frequency and percentage of values present (nonmissing) and the percentage of missing values for each category as related to the other variables.

- No tables are produced if there are no categorical variables.
- Each categorical variable yields a table, whether it is a string variable assumed to be categorical or a numeric variable declared on the CATEGORICAL subcommand.
- The categories of the categorical variable define the columns of the table.
- Each of the remaining variables defines several rows—one each for the number of values present, the percentage of values present, and the percentage of system-missing values; and one each for the percentage of values defined as each discrete type of user-missing (if they are defined).

PERCENT=n   *Omit rows for variables with less than the specified percentage of missing values.* You can specify a percentage from 0 to 100. The default is 5, indicating the omission of any variable with less than 5% missing values. If you specify 0, all rows are displayed.

Example

```
MVA VARIABLES=age income91 childs jazz folk
   /CATEGORICAL=jazz folk
   /CROSSTAB PERCENT=0.
```

- A table of univariate statistics is displayed by default.
- In the output are two crosstabulations, one for *jazz* and one for *folk*. The table for *jazz* displays, for each category of *jazz*, the number and percentage of present values for *age*, *income91*, *childs*, and *folk*. It also displays, for each category of *jazz*, the percentage of each type of missing value (system-missing and user-missing) in the other variables. The second crosstabulation shows similar counts and percentages for each category of *folk*.
- No rows are omitted, since PERCENT=0.

MISMATCH Subcommand

MISMATCH produces a matrix showing percentages of cases for a pair of variables in which one variable has a missing value and the other variable has a nonmissing value (a mismatch). The diagonal elements are percentages of missing values for a single variable, while the off-diagonal elements are the percentage of mismatch of the indicator variables (see "Missing Indicator Variables" on p. 66). Rows and columns are sorted on missing patterns.

PERCENT=n   *Omit patterns involving less than the specified percentage of cases.* You can specify a percentage from 0 to 100. The default is 5, indicating the omission of any pattern found in less than 5% of the cases.

NOSORT   *Suppress sorting of the rows and columns.* The order of the variables in the variables list is used. If ALL was used in the variables list, the order is that of the data file.

DPATTERN Subcommand

DPATTERN lists the missing values and extreme values for each case symbolically. For a list of the symbols used, see "Symbols" on p. 66.

By default, the cases are listed in the order in which they appear in the file. The following keywords are available:

SORT=varname [(order)]   *Sort the cases according to the values of the named variables.* You can specify more than one variable for sorting. Each sort variable can be in ASCENDING or DESCENDING order. The default order is ASCENDING.

DESCRIBE=varlist   *List values of each specified variable for each case.*

Example
```
MVA VARIABLES=populatn density urban religion lifeexpf region
  /CATEGORICAL=region
  /ID=country
  /DPATTERN DESCRIBE=region religion SORT=region.
```
- In the data pattern table, the variables form the columns, and each case, identified by its country, defines a row.
- Missing and extreme values are indicated in the table, and, for each row, the number missing and percentage of variables that have missing values are listed.
- The values of *region* and *religion* are listed at the end of the row for each case.
- The cases are sorted by *region* in ascending order.
- Univariate statistics are displayed.

MPATTERN Subcommand

The MPATTERN subcommand symbolically displays patterns of missing values for cases that have missing values. The variables form the columns. Each case that has any missing values in the specified variables forms a row. The rows are sorted by missing value patterns. For use of symbols, see "Symbols" on p. 66.
- The rows are sorted to minimize the differences between missing patterns of consecutive cases.
- The columns are also sorted according to missing patterns of the variables.

The following keywords are available:

NOSORT *Suppress the sorting of variables.* The order of the variables in the variables list is used. If ALL was used in the variables list, the order is that of the data file.

DESCRIBE=varlist *List values of each specified variable for each case.*

Example
```
MVA VARIABLES=populatn density urban religion lifeexpf region
  /CATEGORICAL=region
  /ID=country
  /MPATTERN DESCRIBE=region religion.
```
- A table of missing data patterns is produced.
- The *region* and the *religion* are named for each case listed.

TPATTERN Subcommand

The TPATTERN subcommand displays a tabulated patterns table, which lists the frequency of each missing value pattern. The variables in the variables list form the columns. Each pattern of missing values forms a row, and the frequency of the pattern is displayed.

- An *X* is used to indicate a missing value.
- The rows are sorted to minimize the differences between missing patterns of consecutive cases.
- The columns are sorted according to missing patterns of the variables.

The following keywords are available:

NOSORT
Suppress the sorting of the columns. The order of the variables in the variables list is used. If ALL was used in the variables list, the order is that of the data file.

DESCRIBE=varlist
Display values of variables for each pattern. Categories for each named categorical variable form columns in which the number of each pattern of missing values is tabulated. For quantitative variables, the mean value is listed for the cases having the pattern.

PERCENT=n
Omit patterns that describe fewer than 1% of the cases. You can specify a percentage from 0 to 100. The default is 1, indicating the omission of any pattern representing less than 1% of the total cases. If you specify 0, all patterns are displayed.

Example

```
MVA VARIABLES=populatn density urban religion lifeexpf region
  /CATEGORICAL=region
  /TPATTERN NOSORT DESCRIBE=populatn region.
```

- Missing value patterns are tabulated. Each row displays a missing value pattern and the number of cases having that pattern.
- DESCRIBE causes the mean value of *populatn* to be listed for each pattern. For the categories in *region*, the frequency distribution is given for the cases having the pattern in each row.

LISTWISE Subcommand

For each quantitative variable in the variables list, the LISTWISE subcommand computes the mean, the covariance between the variables, and the correlation between the variables. The cases used in the computations are listwise nonmissing; that is, they have no missing value in any variable listed in the VARIABLES subcommand.

Example

```
MVA VARIABLES=populatn density urban religion lifeexpf region
  /CATEGORICAL=region
  /LISTWISE.
```

- Means, covariances, and correlations are displayed for *populatn*, *density*, *urban*, and *lifeexpf*. Only cases that have values for all of these variables are used.

PAIRWISE Subcommand

For each pair of quantitative variables, the PAIRWISE subcommand computes the number of pairwise nonmissing values, the pairwise means, the pairwise standard deviations, the pairwise covariance, and the pairwise correlation matrices. These results are organized as matrices. The cases used are all cases having nonmissing values for the pair of variables for which each computation is done.

Example
```
MVA VARIABLES=populatn density urban religion lifeexpf region
  /CATEGORICAL=region
  /PAIRWISE.
```

- Frequencies, means, standard deviations, covariances, and the correlations are displayed for *populatn*, *density*, *urban*, and *lifeexpf*. Each calculation uses all cases that have values for both variables under consideration.

EM Subcommand

The EM subcommand uses an EM (expectation-maximization) algorithm to estimate the means, the covariances, and the Pearson correlations of quantitative variables. This is an iterative process, which uses two steps for each iteration. The E step computes expected values conditional on the observed data and the current estimates of the parameters. The M step calculates maximum likelihood estimates of the parameters based on values computed in the E step.

- If no variables are listed in the EM subcommand, estimates are performed for all quantitative variables in the variables list.
- If you want to limit the estimation to a subset of the variables in the list, specify a subset of quantitative variables to be estimated after the subcommand name EM. You can also list, after the keyword WITH, the quantitative variables to be used in estimating.
- The output includes tables of means, correlations, and covariances.
- The estimation, by default, assumes that the data are normally distributed. However, you can specify a multivariate *t* distribution with a specified number of degrees of freedom or a mixed normal distribution with any mixture proportion (PROPORTION) and any standard deviation ratio (LAMBDA).
- You can save a data file with the missing values filled in. You must specify a filename and its complete path in single or double quotation marks.
- Criteria keywords and OUTFILE specifications must be enclosed in a single pair of parentheses.

The criteria for the EM subcommand are as follows:

TOLERANCE=value *Numerical accuracy control.* The tolerance helps eliminate predictor variables that are highly correlated with other predictor variables and would reduce the accuracy of the matrix inversions involved in the calculations. The smaller the tolerance, the more inaccuracy is tolerated. The default value is 0.001.

CONVERGENCE=value *Convergence criterion.* Determines when iteration ceases. If the relative change in the likelihood function is less than this value, convergence is assumed. The value of this ratio must be between 0 and 1. The default value is 0.0001.

ITERATIONS=n *Maximum number of iterations.* Limits the number of iterations in the EM algorithm. Iteration stops after this many iterations even if the convergence criterion is not satisfied. The default value is 25.

Possible distribution assumptions:

TDF=n *Student's t distribution with n degrees of freedom.* The degrees of freedom must be specified if you use this keyword. The degrees of freedom must be an integer greater than or equal to 2.

LAMBDA=a *Ratio of standard deviations of a mixed normal distribution.* Any positive real number can be specified.

PROPORTION=b *Mixture proportion of two normal distributions.* Any real number between 0 and 1 can specify the mixture proportion of two normal distributions.

The following keyword produces a new data file:

OUTFILE='file' *Specify the name of the file to be saved.* Missing values for predicted variables in the file are filled in by using the EM algorithm. Specify the complete path in single or double quotation marks.

Examples

```
MVA VARIABLES=males to tuition
 /EM (OUTFILE='c:\colleges\emdata.sav').
```

- All variables on the variables list are included in the estimations.
- The output includes the means of the listed variables, a correlation matrix, and a covariance matrix.
- A new data file named *emdata.sav* with imputed values is saved in the *c:\colleges* directory.

```
MVA VARIABLES=all
 /EM males msport WITH males msport gradrate facratio.
```

- For *males* and *msport*, the output includes a vector of means, a correlation matrix, and a covariance matrix.
- The values in the tables are calculated using imputed values for *males* and *msport*. Existing observations for *males*, *msport*, *gradrate*, and *facratio* are used to impute the values that are used to estimate the means, correlations, and covariances.

```
MVA VARIABLES=males to tuition
 /EM verbal math WITH males msport gradrate facratio
   (TDF=3 OUTFILE='c:\colleges\emdata.sav').
```

- The analysis uses a *t* distribution with three degrees of freedom.
- A new data file named *emdata.sav* with imputed values is saved in the *c:\colleges* directory.

REGRESSION Subcommand

The REGRESSION subcommand estimates missing values using multiple linear regression. It can add a random component to the regression estimate. Output includes estimates of means, a covariance matrix, and a correlation matrix of the variables specified as predicted.

- By default, all of the variables specified as predictors (after WITH) are used in the estimation, but you can limit the number of predictors (independent variables) by NPREDICTORS.
- Predicted and predictor variables, if specified, must be quantitative.
- By default, REGRESSION adds the observed residuals of a randomly selected complete case to the regression estimates. However, you can specify that the program add random normal, t, or no variates instead. The normal and t distributions are properly scaled, and the degrees of freedom can be specified for the t distribution.
- If the number of complete cases is less than half the total number of cases, the default ADDTYPE is NORMAL instead of RESIDUAL.
- You can save a data file with the missing values filled in. You must specify a filename and its complete path in single or double quotation marks.
- The criteria and OUTFILE specifications for the REGRESSION subcommand must be enclosed in a single pair of parentheses.

The criteria for the REGRESSION subcommand are as follows:

TOLERANCE=value *Numerical accuracy control.* The tolerance helps eliminate predictor variables that are highly correlated with other predictor variables and would reduce the accuracy of the matrix inversions involved in the calculations. If a variable passes the tolerance criterion, it is eligible for inclusion. The smaller the tolerance, the more inaccuracy is tolerated. The default value is 0.001.

FLIMIT=n *F-to-enter limit.* The minimum value of the F statistic that a variable must achieve in order to enter the regression estimation. You may want to change this limit, depending on the number of variables and the correlation structure of the data. The default value is 4.

NPREDICTORS=n *Maximum number of predictor variables.* This specification limits the total number of predictors in the analysis. The analysis uses the stepwise selected n best predictors, entered in accordance with the tolerance. If $n = 0$, it is equivalent to replacing each variable with its mean.

ADDTYPE *Type of distribution from which the error term is randomly drawn.* Random errors can be added to the regression estimates before the means, correlations, and covariances are calculated. You can specify one of the following types:

RESIDUAL. Error terms are chosen randomly from the observed residuals of complete cases to be added to the regression estimates.

NORMAL. Error terms are randomly drawn from a distribution with the expected value 0 and the standard deviation equal to the square root of the mean squared error term (sometimes called the **root mean squared error**, or RMSE) of the regression.

T(n). Error terms are randomly drawn from the t(n) distribution and scaled by the RMSE. The degrees of freedom can be specified in parentheses. If T is specified without a value, the default degrees of freedom is 5.

NONE. Estimates are made from the regression model with no error term added.

The following keyword produces a new data file:

OUTFILE *Specify the name of the new data file to be saved.* Missing values for the dependent variables in the file are imputed (filled in) by using the regression algorithm. Specify the complete path in single or double quotation marks.

Examples

```
MVA VARIABLES=males to tuition
 /REGRESSION (OUTFILE='c:\colleges\regdata.sav').
```

- All variables in the variables list are included in the estimations.
- The output includes the means of the listed variables, a correlation matrix, and a covariance matrix.
- A new data file named *regdata.sav* with imputed values is saved in the *c:\colleges* directory.

```
MVA VARIABLES=males to tuition
 /REGRESSION males verbal math WITH males verbal math faculty
    (ADDTYPE = T(7)).
```

- The output includes the means of the listed variables, a correlation matrix, and a covariance matrix.
- A *t* distribution with 7 degrees of freedom is used to produce the randomly assigned additions to the estimates.

Bibliography

Azen, S. P., M. Van Guilder, and M. A. Hill. 1989. Estimation of parameters and missing values under a regression model with nonnormally distributed and nonrandomly incomplete data. *Statistics in Medicine*, 8: 217–228.

Dempster, A. P., N. M. Laird, and D. B. Rubin. 1977. Maximum likelihood from incomplete data via the EM algorithm. *Journal of the Royal Statistical Society B: Methodological*, 39: 1–38.

Hill, M. A., and W. J. Dixon. 1981. Missing data: Search for patterns. In *Proceedings of the Statistical Computing Section*, 57–60. American Statistical Association.

Little, R. J. A., and D. B. Rubin. 1987. *Statistical analysis with missing data.* New York: John Wiley and Sons.

Little, R. J. A., and N. Schenker. 1995. Missing data. In *Handbook of Statistical Modeling for the Social and Behavioral Sciences*, G. Arminger, C. C. Clogg, and M. E. Sobel, eds. New York: Plenum Press.

Rubin, D. B. 1987. *Multiple imputation for nonresponse in surveys.* New York: John Wiley and Sons.

Subject Index

chi-square test
  in Missing Value Analysis, 48
correlation estimates
  comparing, 49
correlations
  in Missing Value Analysis, 6, 7, 46
covariance
  in Missing Value Analysis, 6, 7, 46, 49
covariance estimates
  comparing, 49
crosstabulations
  in Missing Value Analysis, 28, 39, 69

data files, 12
data patterns, 17
data sets
  large, 31

EM estimates
  comparing with regression, 56
  in Missing Value Analysis, 7, 41, 73
expectation maximization. See EM estimates
extreme value counts
  in Missing Value Analysis, 5
extreme values
  in Missing Value Analysis, 16, 31, 66

filling in data. See imputation
frequency tables
  in Missing Value Analysis, 5

General Social Survey. See GSS data
GSS data, 12

imputation
  in Missing Value Analysis, 53
  multiple, 59
imputed values, 12
incomplete data. See missing data
indicator variables
  in Missing Value Analysis, 5

listwise deletion
  in Missing Value Analysis, 1
listwise estimation
  in Missing Value Analysis, 41
Little's chi-square test. See chi-square test
Little's MCAR test
  in Missing Value Analysis, 1

MAR test
  in Missing Value Analysis, 42
MATRIX procedue, 49
MCAR test
  in Missing Value Analysis, 1, 42
mean
  in Missing Value Analysis, 5, 6, 7, 16
  pairwise, 43
mismatch
  in Missing Value Analysis, 5, 70
missing at random. See MAR test
missing completely at random. See MCAR test
missing data, 11–59
  casewise patterns, 17
  correlations, 46
  covariance, 46.49
  crosstabulations, 39
  crostbulations, 28
  EM estimation, 41
  estimation, 41
  imputation, 53
  listwise estimation, 41

mismatched patterns, 23
multiple codes, 38
pairwise estimation, 41
pairwise mismatched patterns, 35
regression estimation, 42
summaries, 45
surveys, 29
t tests, 24, 35
univariate statistics, 16, 31
missing data patterns, 14, 19
missing indicator variables
    in Missing Value Analysis, 5, 66
Missing Value Analysis, 1–9, 63–76
    descriptive statistics, 5
    EM, 7
    expectation-maximization, 8
    extreme values, 66
    missing indicator variables, 66
    patterns, 3
    regression, 6
    saving imputed data, 74
    summary tables, 65
    symbols, 66
    univariate statistics, 5
missing value codes, 38
missing value patterns
    in Missing Value Analysis, 70–72
multiple imputation, 59

normal variates
    in Missing Value Analysis, 6

omitting patterns
    in Missing Value Analysis, 3

pairwise estimation
    in Missing Value Analysis, 1, 41
pairwise frequencies
    in Missing Value Analysis, 22
pairwise means
    in Missing Value Analysis, 43
pairwise mismatched patterns
    in Missing Value Analysis, 23, 35

pairwise standard deviations
    in Missing Value Analysis, 45
patterns of missing data, 14

regression estimates
    comparing with EM, 56
    in Missing Value Analysis, 6, 42, 75
residuals
    in Missing Value Analysis, 6

separate variance t tests, 25
sorted casewise patterns
    in Missing Value Analysis, 19
sorting cases
    in Missing Value Analysis, 3
standard deviation
    in Missing Value Analysis, 5, 16, 45
Student's t test
    in Missing Value Analysis, 6
survey data, 29

t tests
    in Missing Value Analysis, 5, 24, 35, 68
    separate variance, 25
tabulated patterns
    in Missing Value Analysis, 20, 33
tabulating cases
    in Missing Value Analysis, 3

univariate statistics
    in Missing Value Analysis, 16, 31

Syntax Index

ADDTYPE (keyword)
  MVA command, 76

CATEGORICAL (subcommand)
  MVA command, 66
CONVERGE (keyword)
  MVA command, 74
COUNTS (keyword)
  MVA command, 69
CROSSTAB (subcommand)
  MVA command, 69

DESCRIBE (keyword)
  MVA command, 70, 71, 72
DF (keyword)
  MVA command, 68
DPATTERN (subcommand)
  MVA command, 70

EM (subcommand)
  MVA command, 73

FLIMIT (keyword)
  MVA command, 75

ID (subcommand)
  MVA command, 67
ITERATIONS (keyword)
  MVA command, 74

LAMBDA (keyword)
  MVA command, 74

LISTWISE (subcommand)
  MVA command, 72

MAXCAT (subcommand)
  MVA command, 67
MEANS (keyword)
  MVA command, 69
MISMATCH (subcommand)
  MVA command, 70
MPATTERN (subcommand)
  MVA command, 71
MVA (command), 63
  CATEGORICAL subcommand, 66
  CROSSTAB subcommand, 69
  DPATTERN subcommand, 70
  EM subcommand, 73
  ID subcommand, 67
  LISTWISE subcommand, 72
  MAXCAT subcommand, 67
  MISMATCH subcommand, 70
  missing indicator variables, 66
  MPATTERN subcommand, 71
  NOUNIVARIATE subcommand, 67
  PAIRWISE subcommand, 73
  REGRESSION subcommand, 75
  symbols, 66
  TPATTERN subcommand, 72
  TTEST subcommand, 68
  VARIABLES subcommand, 66

NOCOUNTS (keyword)
  MVA command, 69
NODF (keyword)
  MVA command, 68
NOMEANS (keyword)
  MVA command, 69
NOPROB (keyword)
  MVA command, 69
NORMAL (keyword)
  MVA command, 76

NOSORT (keyword)
    MVA command, 70, 71, 72
NOT (keyword)
    MVA command, 68
NOUNIVARIATE (subcommand)
    MVA command, 67
NPREDICTORS (keyword)
    MVA command, 75

OUTFILE (keyword)
    MVA command, 74, 76

PAIRWISE (subcommand)
    MVA command, 73
PERCENT (keyword)
    MVA command, 68, 70
PROB (keyword)
    MVA command, 69
PROPORTION (keyword)
    MVA command, 74

REGRESSION (subcommand)
    MVA command, 75
RESIDUAL (keyword)
    MVA command, 76

SORT (keyword)
    MVA command, 70

T (keyword)
    MVA command, 68, 76
TDF (keyword)
    MVA command, 74
TOLERANCE (keyword)
    MVA command, 73, 75
TPATTERN (subcommand)
    MVA command, 72
TTEST (subcommand)
    MVA command, 68
VARIABLES (subcommand)
    MVA command, 66